Sustainable Civil Infrastructures

About this Series

Sustainable Infrastructure impacts our well-being and day-to-day lives. The infrastructures we are building today will shape our lives tomorrow. The complex and diverse nature of the impacts due to weather extremes on transportation and civil infrastructures can be seen in our roadways, bridges, and buildings. Extreme summer temperatures, droughts, flash floods, and rising numbers of freeze-thaw cycles pose challenges for civil infrastructure and can endanger public safety. We constantly hear how civil infrastructures need constant attention, preservation, and upgrading. Such improvements and developments would obviously benefit from our desired book series that provide sustainable engineering materials and designs. The economic impact is huge and much research has been conducted worldwide. The future holds many opportunities, not only for researchers in a given country, but also for the worldwide field engineers who apply and implement these technologies. We believe that no approach can succeed if it does not unite the efforts of various engineering disciplines from all over the world under one umbrella to offer a beacon of modern solutions to the global infrastructure. Experts from the various engineering disciplines around the globe will participate in this series, including: Geotechnical, Geological, Geoscience, Petroleum, Structural, Transportation, Bridge, Infrastructure, Energy, Architectural, Chemical and Materials, and other related Engineering disciplines.

More information about this series at http://www.springer.com/series/15140

Shuying Wang · Yu Xinbao
Moses Tefe
Editors

New Solutions for Challenges in Applications of New Materials and Geotechnical Issues

Proceedings of the 5th GeoChina International Conference 2018 – Civil Infrastructures Confronting Severe Weathers and Climate Changes: From Failure to Sustainability, held on July 23 to 25, 2018 in HangZhou, China

Springer

Editors
Shuying Wang
School of Civil Engineering
Central South University
Changsha, China

Moses Tefe
Norwich University
Northfield, VT, USA

Yu Xinbao
Department of Civil Engineering
The University of Texas at Arlington
Arlington, TX, USA

ISSN 2366-3405 ISSN 2366-3413 (electronic)
Sustainable Civil Infrastructures
ISBN 978-3-319-95743-2 ISBN 978-3-319-95744-9 (eBook)
https://doi.org/10.1007/978-3-319-95744-9

Library of Congress Control Number: 2018948649

This Springer imprint is published by the registered company Springer Nature Switzerland AG
The registered company address is: Gewerbestrasse 11, 6330 Cham, Switzerland

Contents

Introduction

Although geotechnical engineering is a traditional research field, it still bears many great challenges, which scientists and engineers commonly face. To make the civil infrastructures sustainable, new technologies and materials have been proposed and applied in the field. This volume comprises a set of high-quality, refereed papers. They address the different aspects of new solutions for challenges in geotechnical engineering, such as characterization and applications of new materials (e.g. natural jute fibres, geotextile fabric and permeable concrete), modified constitutive models, new numerical technologies for traditional problems, and some other geotechnical issues that are becoming quite relevant in today's world practice. This volume is a part of the Proceedings of the 5th GeoChina International Conference 2018—Civil Infrastructures Confronting Severe Weathers and Climate Changes: From Failure to Sustainability, HangZhou, China.

Modified Duncan-Chang Model and Mechanics Parameter Determination Based on Triaxial Consolidated Drained Tests of Guiyang Red Clay in China

Yongxin Ji[1], Shuangying Zuo[2(✉)], Jing Zhang[2], and Yanzhao Zhang[2]

[1] Traffic Science Department, Guizhou Construction Science Research and Design Institute of CSCEC, Nanming District, Guiyang 550006, China
574283019@qq.com

[2] College of Resources and Environment Engineering, Guizhou University, Huaxi District, Guiyang 550025, China
shyzuo@163.com, 1464750356@qq.com, 1533837365@qq.com

Abstract. A large area of China Southwest is covered by red clay, which is a kind of Karst regional special clay with high moisture content, high ductility, large void ratio, low density, poor compaction. Based on Duncan-Chang model, the constitutive model of red clay and its mechanical parameters are studied by laboratory tests and numerical simulation. (1) According to the results of triaxial consolidated drained tests with different moisture content, the damage forms, stress-strain curve, the relationships between pore water pressure and displacement changing with the time are analyzed in detail. (2) Based on Duncan-Chang model, the incremental elastic curve is fitted. The curve is composed of three segments which represent the deformation law of the clay samples in compression-shearing process. Correspondingly, the initial tangent elastic modulus is modified into a three-stage function, and the Duncan-Chang model of Guiyang red clay is obtained and the parameters such as shear strength, failure ratio, tangent Poisson's ratio and tangent volume modulus are calculated. (3) VC ++ programming language is used to modify Duncan-Chang E-B model in the software of $FLAC^{3D}$. The results show that the numerical simulation curve is in good agreement with the stress-strain curve of the experiments, which shows that the modified Duncan-Chang model has good feasibility and applicability to Guiyang red clay.

1 Introduction

Red clay mainly exists in the Yunnan-Guizhou Plateau and its adjacent, central and western regions of Guangxi, western Hunan and other places in China, which is the product of carbonate rock going through physical weathering, chemical weathering and laterization in the condition of humid and warm climate, as well as abundant rainfall. The red clay is a kind of brownish red or isabellinus special soil. It has poorer physical properties of high moisture content, high plasticity, large porosity ratio, but it has relatively good mechanical properties such as high strength, low compressibility and so

S. Wang et al. (eds.), *New Solutions for Challenges in Applications of New Materials and Geotechnical Issues*, Sustainable Civil Infrastructures,
https://doi.org/10.1007/978-3-319-95744-9_1

on. So it can not use the relevant laws of physical indexes and mechanical parameters of general cohesive soil to evaluate the engineering properties of red clay (Liao and Zhu 2004). At the same time, the red clay has the characteristics of typical reverse profile of "upper-hard and lower-soft" and strong water sensitivity such as softening with water absorption and cracking with water loss. To analyze accurately geotechnical engineering problems, it is most important to establish a proper constitutive model. In the past decades, domestic and foreign scholars have put forward many constitutive models on the laws of soil deformation, but in fact, these models are only presented for saturated disturbed soils or sand soils (Shen 1996).

The most famous constitutive models of soils are Duncan-Chang model, Cambridge model, K-G model, Zhujiang Shen model, Tsinghua model and so on. However, the classic constitutive models of red clay have not yet proposed. Based on the stress-strain relationship of Chongqing red clay in different stress paths, Zhang and Deng (1997) considered that K-G model and the elastic-plastic model of Yin's double-yield surface can reflect constitutive relationship of Chongqing red clay, and the corresponding parameters were provided. Liu (2009) used the Mohr-Coulomb model and the Drucker-prager plastic model respectively to simulate the foundation settlement of different engineering positions of the Wuhan-Guangzhou express railway in the $FLAC^{3D}$ software, which found that the calculation error was relatively small and proved that the numerical simulation is a more effective method for predicting the settlement value during embankment fills and after construction. On the basis of the previous research results, Chen and Xing (2010) set up the stress-strain relationship curve of Guiyang's red clay with the power function and established loading constitutive model and humidifying (or dehumidifying) constitutive model with high liquid limit. Based on a large number of indoor dynamic triaxial tests, Liu et al. (2011) studied the dynamic constitutive relationship and dynamic modulus attenuation law of undisturbed red clay in different experimental conditions. Zhou et al. (2014), Fang (2011) used the Mohr-Coulomb model to simulate the excavation process of red clay in deep foundation pit and the stability of anchor reinforcement in bedding red clay slope in $FLAC^{3D}$ software respectively.

Decades of researches for red clay have made a lot of achievements, but there still exist some disadvantages in the mechanical strength, calculation model and numerical simulation application. In this article, based on triaxial consolidated drained shear tests of Guiyang red clay, the Duncan-Chang model is modified and developed in FLAC3D software, and the numerical experiment is carried out to verify the applicability of the model.

2 Triaxial Consolidated Drained Test of Red Clay

2.1 Preparation of Standard Samples

Take some soil samples from a deep foundation pit in Guiyang, and the soil samples are cut into standard cylinders of 39.1 mm in diameter and 80 mm in height according to the Geotechnical Test Protocol (SL237-1999). 28 samples are prepared at one time and the initial moisture content is 57.03%. In order to prepare four groups of samples with different moisture contents, 28 samples were divided into 4 groups at random,

7 samples in each group. One group keeps their initial moisture content, other three groups are air-dried for 1 day, 2 days and 3 days respectively, then measuring the moisture content of each group is 54.5%, 52%, 47%. In this case, swelling-shrinkage potential of red clay during the air-dried process can be ignored.

2.2 Experiment Methods

The TSZ-1 strain-controlled triaxial instrument produced by Nanjing Soil Instrument Factory is used for the consolidated and drained shear test. In order to study the mechanical properties of undisturbed red clay with different moisture contents under different stress conditions, the confining pressures of the four groups of samples with different moisture contents are 100 kPa, 150 kPa, 200 kPa, 250 kPa, 350 kPa, 550 kPa and 750 kPa respectively, and the loading velocity is 0.006 mm/min. The samples at any stage did not produce pore water pressure in the course of Consolidated Drained (CD) Test, that is, the total stress is equal to the effective stress, and the volume change of the samples can be measured according to the water discharge. In the test, the data record is cut until the axial strain of soil samples reaches 20%.

2.3 Analysis of Experiment Results

Samples damage pattern

In the triaxial CD test, with the increasing load, the radial deformation and failure surface gradually forms, 28 samples all produce significant shear surfaces. Although the sample shows a single mode of shear failure, there are some tiny cracks around the main destruction surface and forming shear zones, and each shear zone has different shapes and patterns, as shown in Fig. 1: red lines are main damage surfaces, yellow lines are small cracks.

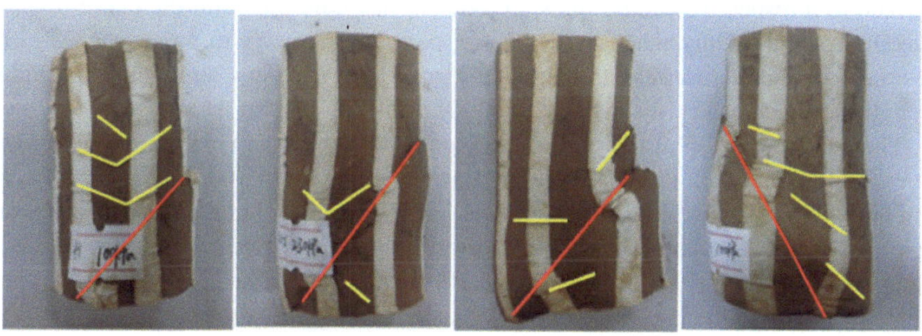

Fig. 1. The destruction patterns of shear zone in triaxial test (Color figure online)

Analysis of stress-strain curve

The test data of the original record are extracted at a certain time interval, and we can obtain $\sigma - \varepsilon$ curves in different confining pressures at the certain moisture content as shown in Fig. 2. For the same moisture content, when the confining pressure increases

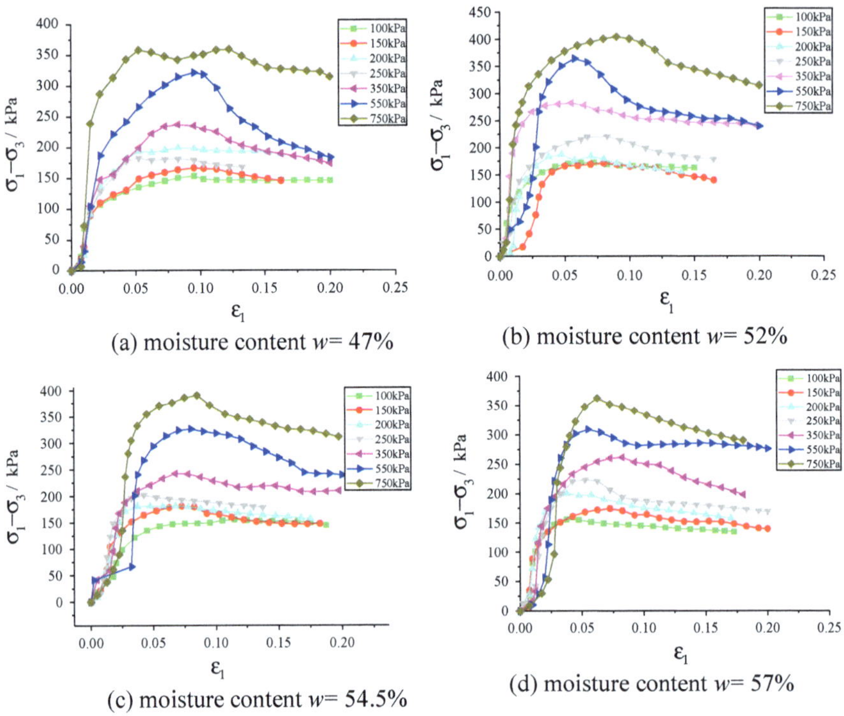

Fig. 2. $(\sigma_1 - \sigma_3) \sim \varepsilon_1$ Curves under changing confining pressure for the same moisture content

from 100 kPa to 550 kPa, the softening property of red clay is gradually increasing, then when the confining pressure increases to 750 kPa, the softening property is gradually weakening. When the confining pressure is at lower stage (100 kPa, 150 kPa and 200 kPa respectively), the moisture content has little effect on the curve $\sigma - \varepsilon$ of Guiyang red clay. Only when the confining pressure increases to more than 250 kPa, the effect of moisture content is gradually manifest, the peak strength and residual strength both decrease with the increasing of moisture content.

The $\sigma - \varepsilon$ curves in different moisture contents for a certain confining pressure are shown in Fig. 3. When the confining pressure increases from 100 kPa to 550 kPa, the softening performance of red clay gradually is more obvious, but the confining pressure increases to 750 kPa, the softening performance gradually decreases. When the confining pressure is at a lower stage (e.g.: 100 kPa, 150 kPa and 200 kPa), the water content has little influence on the curve. When confining pressure increases to 250 kPa, the influence of water content on the curve of red clay is emerging, and with the increasing of water content, the peak strength and residual strength are decreased.

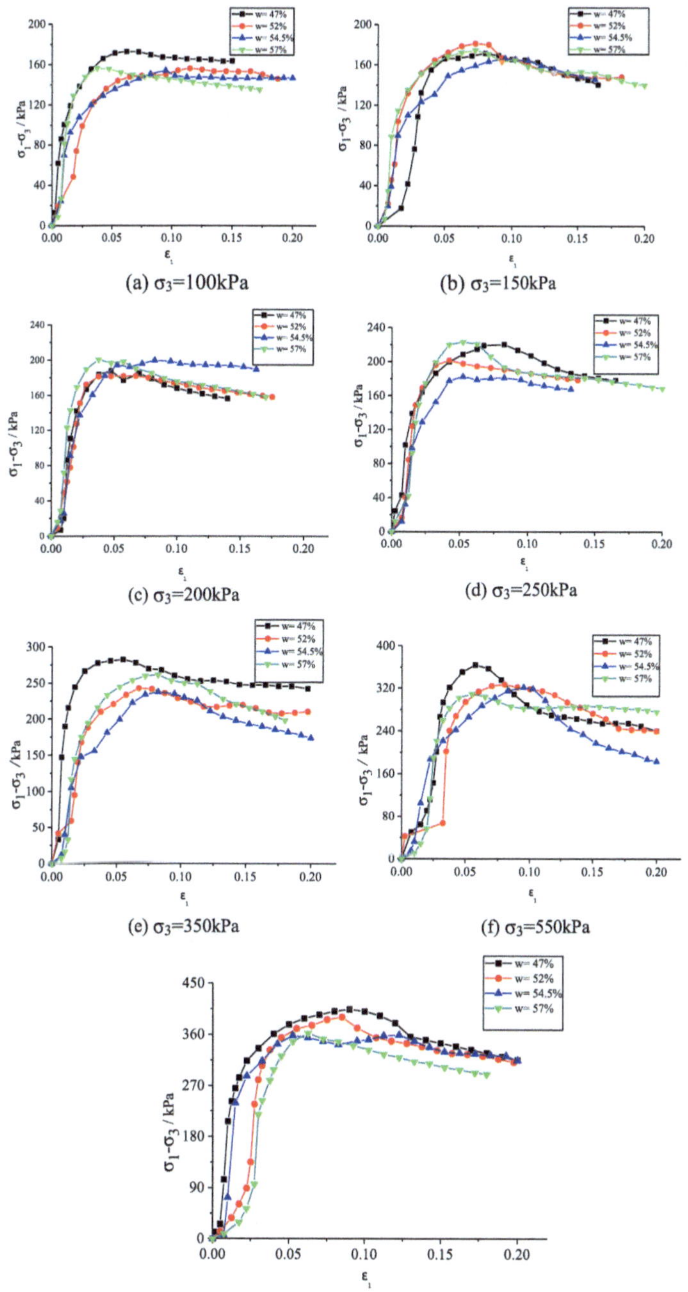

(a) σ_3=100kPa (b) σ_3=150kPa

(c) σ_3=200kPa (d) σ_3=250kPa

(e) σ_3=350kPa (f) σ_3=550kPa

(g) σ_3=750kPa

Fig. 3. $(\sigma_1 - \sigma_3) \sim \varepsilon_1$ Curves under changing moisture content for the same confining pressure

Analysis of pore water pressure

The relationships between the ultra-static pore pressure and time in four groups of CD tests are shown in Fig. 4. The dissipation rate of pore pressure decreases exponentially with time. The process of dissipation is roughly divided into three stages. When the time is about 0–100 min, the dissipation is the fastest, when the time is about 100–400 min, the dissipation is slower, when the time is over 400 min, the dissipation has a steady trend.

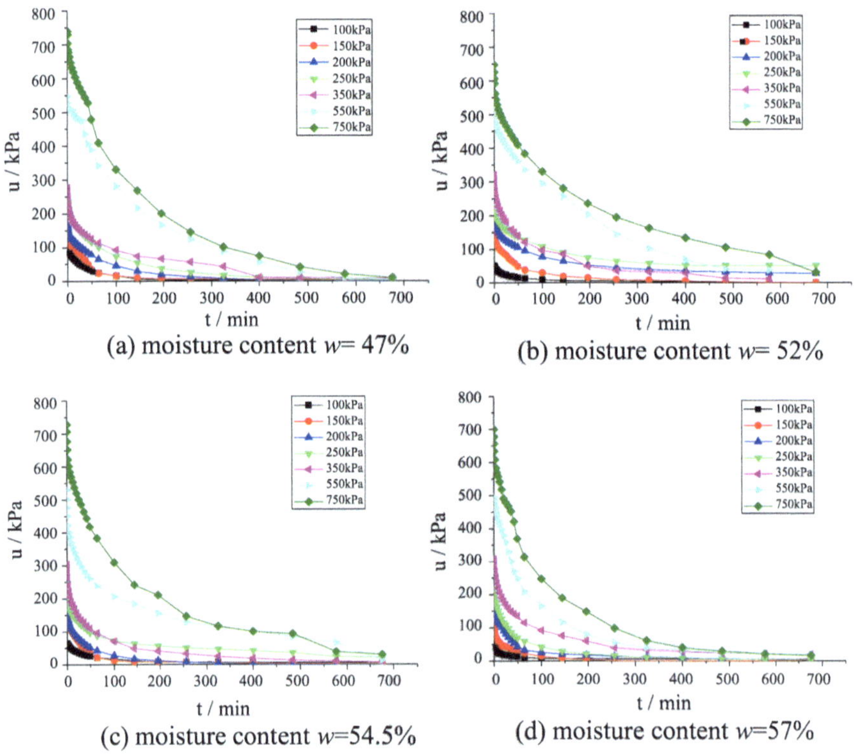

Fig. 4. Curves between pore water pressure and time

Analysis of water discharge

The magnitude of water discharge is equal to the change of soil volume during the compression process, which can help us to determine quantitatively the degree of consolidation. The water discharge variation from the soil with time is shown in Fig. 5. The relationship curve of the discharge and the time is basically hyperbolic shape. In the initial state of consolidation, that is, about the first 100 min, the drainage is rapid, the discharge is large, and then keeps in a stable state after 100 min. In addition to slight fluctuations caused by local heterogeneity of samples, the same rules are observed: the greater the confining pressure, the faster the drained rate, the greater the discharge, the more the soil volume changes.

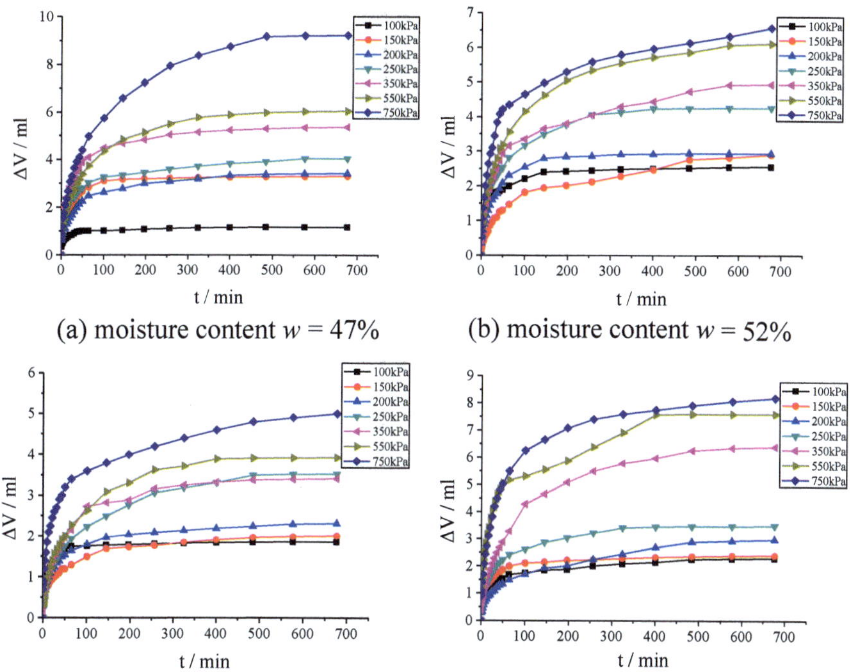

(a) moisture content $w = 47\%$
(b) moisture content $w = 52\%$
(c) moisture content $w = 54.5\%$
(d) moisture content $w = 57\%$

Fig. 5. $\Delta V - t$ curves with different moisture content

3 Modified Duncan-Chang Model of Guiyang Red Clay

3.1 Duncan-Chang Model

Kondner (1963) proposed that the triaxial test curve $(\sigma_1 - \sigma_3) \sim \varepsilon_a$ of the general soil could be fitted with hyperbola in accordance with a large number of soil triaxial tests (Fig. 6).

$$\sigma_1 - \sigma_3 = \frac{\varepsilon_a}{a + b\varepsilon_a} \tag{1}$$

Where, a and b are test constants, for conventional triaxial compression tests, $\varepsilon_a = \varepsilon_1$. Duncan et al. presented a widely used incremental elastic model based on this hyperbolic stress-strain relationship, which is generally called the Duncan-Chang model.

The results of the conventional triaxial compression test are settled in accordance with the relationship of $\varepsilon_1/(\sigma_1 - \sigma_3) \sim \varepsilon_1$, the curve is approximately linear relationship (Fig. 7). Where, a, b is the intercept and slope of the straight line respectively.

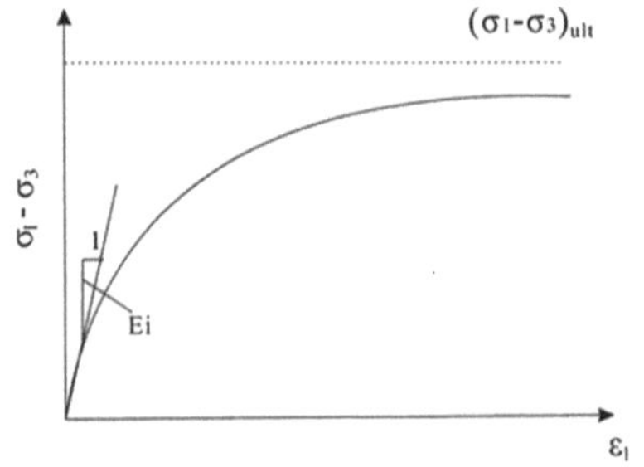

Fig. 6. $(\sigma_1 - \sigma_3) \sim \varepsilon_1$ hyperbola

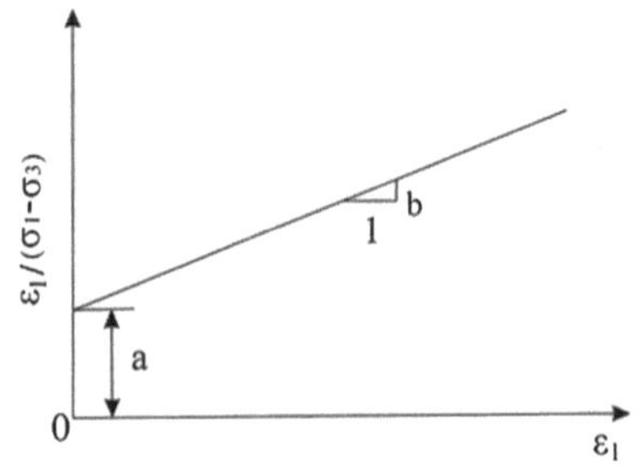

Fig. 7. $\varepsilon_1/(\sigma_1 - \sigma_3) \sim \varepsilon_1$ Line

Tangent deformation modulus
Because of $d\sigma_2 = d\sigma_3 = 0$, tangent modulus is given as

$$E_t = \frac{d(\sigma_1 - \sigma_3)}{d\varepsilon_1} = \frac{a}{(a + b\varepsilon_1)^2} \tag{2}$$

When $\varepsilon_1 = 0$, $E_T = E_I$ then $E_I = 1/a$; the definition of damage ratio is

$$R_f = \frac{(\sigma_1 - \sigma_3)_f}{(\sigma_1 - \sigma_3)_{ult}}, \text{ then } b = \frac{1}{(\sigma_1 - \sigma_3)_{ult}} = \frac{R_f}{(\sigma_1 - \sigma_3)_f}$$

The relationship of $\lg(E_i/p_a)$ and $\lg(\sigma_3/p_a)$ is almost a straight line, so Eq. (3) is available as

$$E_i = Kp_a\left(\frac{\sigma_3}{p_a}\right)^n \tag{3}$$

Where, P_A is the atmospheric pressure ($\mathrm{p_a} = 101.4\ \mathrm{kP_a}$), the dimension and unit is the same as σ_3; K and N are test constants, representing the intercept and the slope of the straight line respectively.

According to the Mohr-Coulomb strength criterion, there are

$$(\sigma_1 - \sigma_3)_f = \frac{2c\cos\varphi + 2\sigma_3 \sin\varphi}{1 - \sin\varphi} \tag{4}$$

Substituting Eqs. (3) and (4) into Eq. (2) respectively, there are

$$E_t = Kp_a\left(\frac{\sigma_3}{p_a}\right)^n \left[1 - \frac{R_f(\sigma_1 - \sigma_3)(1 - \sin\varphi)}{2c\cos\varphi + 2\sigma_3\sin\varphi}\right]^2 \tag{5}$$

Volume modulus

Duncan et al. (1980) presented *E-B* model, in which the bulk modulus *B* replaced tangent Poisson's ratio, i.e.

$$B = \frac{E_t}{3(1 - 2v_i)} \tag{6}$$

The change of average normal stress is $\Delta p = (\sigma_1 - \sigma_3)/3$ as applying partial stress in the process of triaxial test, therefore

$$B_t = \frac{1}{3}\frac{\partial(\sigma_1 - \sigma_3)}{\partial \varepsilon_v} \tag{7}$$

Duncan et al. assumed that *B* is independent of the stress level *s* and the principal stress difference $(\sigma_1 - \sigma_3)$. For the same σ_3, if the relationship curve of $(\sigma_1 - \sigma_3)/3 \sim \varepsilon_v$ is plotted, regarding the slope of the line between the dot corresponding to the stress level *s* = 0.7 and origin as the average slope, i.e.

$$B = \frac{(\sigma_1 - \sigma_3)_{s=0.7}}{3(\varepsilon_{vs=0.7})} \tag{8}$$

Where, $(\sigma_1 - \sigma_3)_{70\%}$ and $\varepsilon_{v70\%}$ are the deviation stress and the volumetric strain respectively when the value of $(\sigma_1 - \sigma_3)$ reaches for 70% of $(\sigma_1 - \sigma_3)_f$. Thus, *B* is a constant for each of triaxial compression tests when σ_3 is a constant. In the double logarithmic coordinate, the relationship curve between $\lg(B_t/p_a)$ and $\lg(\sigma_3/p_a)$ is plotted with dots, we can get a straight line whose intercept and slope is K_b and *m*, then

$$B_t = K_b p_a \left(\frac{\sigma_3}{p_a}\right)^m \tag{9}$$

As the value of *v* can only change in the range of 0–0.49, the value of B_T must be restricted in the range of (0.33–17)E_T.

3.2 Modification Method of Duncan-Chang Model

Deformation mechanism of red clay

During the consolidation process, the confining pressure is not sufficient to destroy the soil, and Nagaraj (1990) proposed the stress-strain curve of the natural structural soil was divided into three sections during the shearing process. In the first stage, the difference of main stress $(\sigma_1 - \sigma_3)$ gradually increases with the increasing of axial strain. The soil structure maintains an intact state, basically in the elastic deformation stage. The second stage is that when the difference of main stress increases to the peak. A lot of destruction of soil begins to occur, and the obvious shear surface appears, which suggests the soil has broken the yield point and destroyed its original structure, At this point, the structure of the soil has produced a lot of destruction, soil particles slip each other, and accompanied by a large number of structural collapse. In the third stage, when the deformation of the soil continues to increase until the structure of the soil is completely destroyed, the load that the soil can bear decreases with the aggravating of the soil deformation, and the difference of main stress $(\sigma_1 - \sigma_3)$ is stable, which suggests that the soil is completely destroyed. In the process of soil deformation, the interaction between the plastic body strain and the plastic shear strain always exists, that is the reason why the soil has strain softening characteristic (He and Kong 2010). The action of plasticity strain on the shear strain is applied by changing the capacity of shearing. Under shear strain, when the dilatancy condition is satisfied, the volume expands and the shear resistance decreases, then there is a strain softening curve with a hump.

Modified Duncan-Chang model of Guiyang red clay

The $\sigma - \varepsilon$ curve can be directly obtained through the conventional triaxial test, and it is rewritten as $\varepsilon_1/(\sigma_1 - \sigma_3) \sim \varepsilon_1$ to represent the process of soil deformation (Wang et al. 2004), As shown in Fig. 8 (regarding a curve that moisture content is 52% and confining pressure is 100 kPa as an example), the three stages of the curve can be three straight lines to fit: ① $y = 0.00288x + 0.00007$, ② $y = -0.0036x + 0.0005$, ③ $y = 0.0064x + 0.0001$. The reciprocal of the intercept of the three straight lines is the initial tangent modulus, and the slope is $1/(\sigma_1 - \sigma_3)_{ult}$.

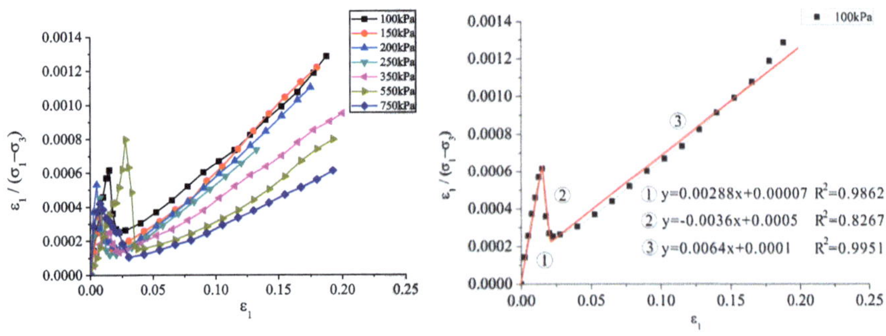

Fig. 8. The $\varepsilon_1/(\sigma_1 - \sigma_3) \sim \varepsilon_1$ curve and fitted lines of three stage (w = 52%, σ_3 = 100 kPa)

According to Eq. (5), the tangent elastic modulus E_t can also be expressed by a three-step function, as shown in Eq. (10), which is the expression of the modified Duncan-Chang model.

$$\begin{cases} E_{t1} = K_1 p_a \left(\frac{\sigma_3}{p_a}\right)^{n_1} \left(1 - \frac{R_f(\sigma_1-\sigma_3)(1-\sin\varphi)}{2c\cos\varphi + 2\sigma_3\sin\varphi}\right)^2 & \text{structural integrity stage} \\ E_{t2} = K_2 p_a \left(\frac{\sigma_3}{p_a}\right)^{n_2} \left(1 - \frac{R_f(\sigma_1-\sigma_3)(1-\sin\varphi)}{2c\cos\varphi + 2\sigma_3\sin\varphi}\right)^2 & \text{structural damage stage} \\ E_{t3} = K_3 p_a \left(\frac{\sigma_3}{p_a}\right)^{n_3} \left(1 - \frac{R_f(\sigma_1-\sigma_3)(1-\sin\varphi)}{2c\cos\varphi + 2\sigma_3\sin\varphi}\right)^2 & \text{structural failure stage} \end{cases} \quad (10)$$

3.2.1 Determination of Parameters

(1) Determination of shear strength parameters

According to a simple method that was proposed by Chen (2003) to obtain cohesion and internal friction angle, that is, the test results will be plotted on the $p-q$ axis, where $p = (\sigma_1 + \sigma_3)/2$, $q = (\sigma_1 - \sigma_3)/2$, and then the test results are fitted with a straight line, if the intercept and slope of the straight line are b_0 and $\tan\alpha$ respectively, the friction angle α and cohesion c are calculated via Eq. (11), the results are shown in Table 1.

Table 1. Shear strength parameters of red clay with different moisture content w

w	b	tan α	φ (°)	cos φ	c (kPa)
47%	51.69	0.1663	9.57	0.9861	52.42
52%	49.40	0.1557	8.96	0.9878	50.01
54.50%	50.33	0.1454	8.36	0.9894	50.87
57%	58.84	0.1359	7.81	0.9907	59.39

$$\varphi = \sin^{-1}(\tan\alpha) \quad c = \frac{b_0}{\cos\varphi} \quad (11)$$

(2) Initial tangent modulus

The various stages of soil structure damage are linearly fitted. The initial tangent modulus *Ei* of each stage and the asymptotic value of the principal stress difference are calculated in accordance with Eq. (2). The calculated results are listed in Tables 2, 3 and 4.

(3) Destruction ratio R_f

The destruction ratio R_f of the three stages is calculated by Eq. (2). The average value under each confining pressure is taken as the determined value, as shown in Table 5.

Table 2. The corresponding E_{i1} values for each confining pressure (unit: kPa)

w	σ_3 (kPa)						
	100	150	200	250	350	550	750
47%	10000	10000	50000	33333	50000	100000	125000
52%	14286	25000	25000	33333	50000	50000	100000
54.5%	12500	25000	25000	50000	50000	100000	125000
57%	14286	16667	25000	25000	33333	50000	50000

Table 3. The corresponding E_{i2} values for each confining pressure (unit: kPa)

w	σ_3 (kPa)						
	100	150	200	250	350	550	750
47%	1250	1250	833	533	417	370	333
52%	2000	1667	2000	1667	2000	1322	1000
54.5%	2500	3333	2500	625	1667	294	400
57%	1000	833	1100	833	667	714	769

Table 4. The corresponding E_{i3} values for each confining pressure (unit: kPa)

w	σ_3 (kPa)						
	100	150	200	250	350	550	750
47%	12500	14286	20000	33333	33333	50000	100000
52%	10000	20000	20000	50000	33333	33333	50000
54.5%	20000	25000	33333	25000	33333	50000	50000
57%	20000	20000	25000	20000	25000	33333	50000

Table 5. Each stage of destruction ratio value R_f

w	R_f		
	R_{f1}	R_{f2}	R_{f3}
47%	0.61	−0.76	0.82
52%	0.60	−0.68	0.83
54.5%	0.62	−0.66	0.84
57%	0.64	−0.72	0.81

(4) Parameters K and n

The relationship diagrams of $\lg(E_i/p_a) \sim \lg(\sigma_3/p_a)$ are plotted in each moisture content respectively, the intercept of fitting line is lg$K$, slope is n, the values of the three stages are shown in Table 6.

(5) Parameters of volume modulus

According to Eqs. (8) and (9), the volume modulus parameters K_b and m are calculated, the results are listed in Table 7.

Table 6. Elastic modulus parameters of Guiyang red clay

w	K_1	K_2	K_3	n_1	n_2	n_3
47%	100.879	12.891	109.194	1.319	−0.751	0.999
52%	151.635	21.306	139.123	0.857	−0.287	0.667
54.5%	137.848	35.653	201.280	1.126	−1.161	0.461
57%	140.055	9.750	170.765	0.679	−0.141	0.421

Table 7. Volumetric modulus parameters K_b and m of Guiyang red clay

w	K_b	m
47%	266.809	0.6081
52%	266.625	0.5739
54.5%	279.963	0.6202
57%	236.157	0.9914

At this time, the eight parameters required by the development of modified Duncan-Chang *E-B* model have all be calculated, they are the shear strength c and φ; the tangential elastic modulus parameters E_{i1}, E_{i2}, E_{i3}, R_{f1}, R_{f2}, R_{f3}, K_1, K_2, K_3, n_1, n_2 and n_3; volume modulus parameters K_b and m.

4 Numerical Test

4.1 Standard Cylindrical Specimen

Triaxial numerical test is used to verify the modified Duncan-Chang model, then the dynamic link library file (ReDuncan.DLL) is embedded into the main program of FLAC^{3D}. The size of the model is the same as laboratory test, the method of loading is vertical axial bi-directional loading, loading rate is $1*10^{-7}$ m/s. According to different confining pressures, the number of calculating steps is from 2000 to 15000 steps. The model sample is shown in Fig. 9. The data with 52% moisture content are taken as examples, calculation parameters are shown in Table 8.

4.2 Analysis of Numerical Test Results

Take soil samples of 52% moisture content as examples. The simulated curves of modified model are compared with the laboratory test curves, and the results are shown in Fig. 10.

It can be seen from the Fig. 10 that the numerical simulated results of FLAC^{3D} and the experimental results have some difference, the peak of the simulation is higher than the peak of the test. When confining pressure is lower, the simulated curve don't show the property of softening after reaching the yield, the peak strength is almost unchanged, and the property of softening is not reflected. When confining pressure is higher, the softening property of the simulated curve is stronger than that of the test curve, there are a significant peak and a residual strength at earlier stage. The simulated

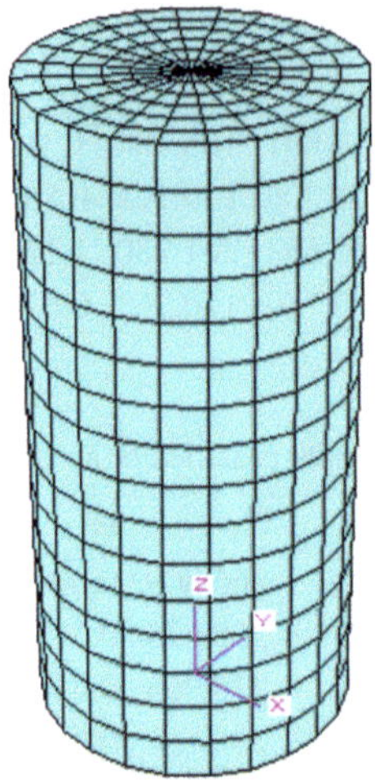

Fig. 9. Standard cylindrical specimen model

Table 8. Calculation parameters of modified Duncan-Chang model

Para-meters	C (kPa)	φ (°)	Rf_1	Rf_2	Rf_3	K_1	K_2	K_3	n_1	n_2	n_3	K_b	m
Value	50	8.96	0.6	−0.68	0.87	152	21	139	0.86	−0.29	0.67	267	0.57

curve is more volatile. Generally, although the simulation curve and the test curve have some differences, they still maintain a good consistency.

4.3 Discussion

1. Due to the big pore size and low compressibility of Guiyang red clay, the stress-strain curve shows the characteristics of weak hardening - weak softening in the process of confining pressure increasing. In order to describe the stress change at the peak point, a three-phase modification of Duncan-Chang model is used to simulate the process. The middle part is the opposite of the first (elastic stage) and the third (failure stage), which indicates that the strength is decreasing after the damage of the soil sample.
2. In the low-middle confining pressure phase, the stress-strain curve shows weak hardening characteristics, which is close to the original Duncan-Chang model. During the stress iteration in the plastic stage, the calculated stress increment is small and the curve is stable and smooth. The numerical test curve is consistent with the test curve.
3. In the high confining pressure phase, the stress increases fast and the peak value appears in advance. In the second stage, the stress increment in the negative direction is greatly modified, and the fluctuation phenomenon may occur during calculation. Therefore, the softening characteristic of the numerical simulation curve is higher than the test curve.

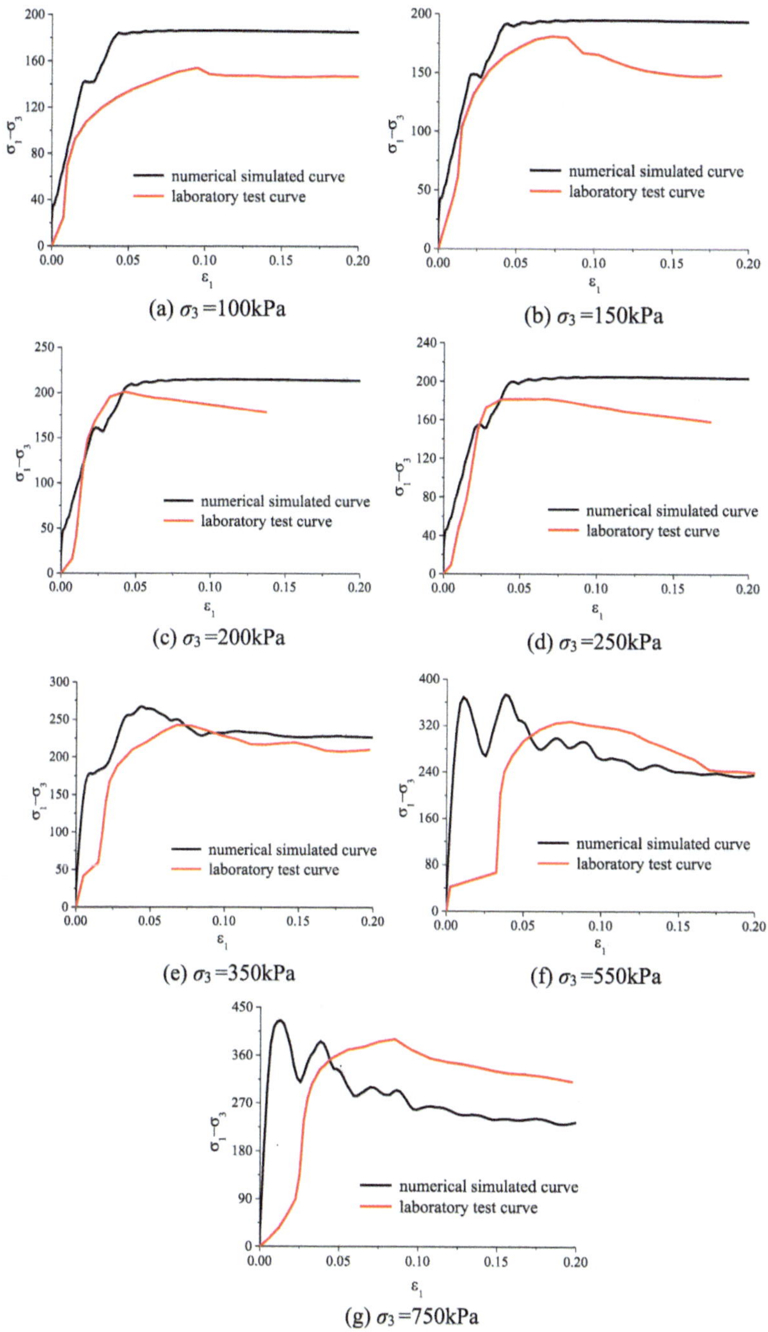

Fig. 10. Curves of comparison of numerical simulation and laboratory test

4.4 Conclusions

On the basis of triaxial CD test, the modified Duncan-Chang model of Guiyang red clay is established by studying the stress-strain relationship of laboratory tests. The model parameters are calculated and the secondary development in $FLAC^{3D}$ is carried out to verify the reasonability of the model.

1. The results of triaxial CD test show that the red clay in Guiyang has obvious shear surface during the shear process, and the stress-strain curve tends to be stable after reaching the obvious peak point. The process from the elastic deformation stage to the plastic deformation stage has a weak softening trend. The pore water pressure in the experiments is exponentially related to the time, and the relationship of water discharge and the time are hyperbola.
2. After fitting the $\varepsilon_1/(\sigma_1 - \sigma_3) \sim \varepsilon_1$ curve, it is found that the curve has obvious three-step linear function, and the initial tangent elastic modulus is modified into segment function, then the Duncan-Chang model of Guiyang's red clay is obtained, and the eight parameters needed for the model are deduced. The modified model can better reflect the development of stress-strain relationship in different stages.
3. After the $FLAC^{3D}$ numerical test, it is found that the trend of simulation curves are in agreement with the stress-strain curves of the laboratory experiments, however there is a little difference in value.
4. The suitability of the modified model for red clay with high confining pressure has yet to be further demonstrated.

Acknowledgments. This work is supported by the National Natural Science Foundation of China (51469007), Guizhou Science and Technology Support Project for Social development ([2015]3055), Science and technology research and development project of CSCEC (CSCEC4B-2015-KT-03). They are gratefully acknowledged.

References

Liao, Y., Zhu, L.: The red clay of Guizhou carbonate rock. Guizhou People Press, Guiyang (2004)

Shen, Z.: Mathematical model of soil structure. J. Geotech. Eng. **18**(1), 95–97 (1996)

Zhang, Y., Deng, A.: Proof test for constitutive model of Chongqing's Laterite. J. Chongqing Jianzhu Univ. **19**(2), 48–53 (1997)

Liu, D.: Research on Calculation and Prediction Method of Red Clay Foundation Settlement in Wuhan-Guangzhou Passenger Dedicated Line. Central South University (2009)

Chen, K., Xing, H.: Reserach on deformation characteristics of high liquid limit red clay. J. Highw. Transp. Res. Dev. **23**(3), 49–54 (2010)

Liu, X., Yang, G., Fang, W.: Dynamic constitutive relation and dynamic modulus attenuation model of red clay. Hydrogeol. Eng. Geol. **38**(3), 66–72 (2011)

Zhou, Y., Liu, X., Yu, Y., et al.: Lateral unloading true triaxial test and analysis of mechanical properties for red clay. J. Chongqing Univ. **37**(6), 18–24 (2014)

Fang, W.: Research on Stability Analysis Method of Cutting Slope of Residual Red Clay. Central South University (2011)

Geotechnical Test Procedures (1999) China Water Conservancy and Hydropower Press, Beijing

Nagaraj, T.S., et al.: Analysis of compressibility of sensitive soils. J. Geoetech. Eng. **116**(l), 105–118 (1990)

He, L., Kong, L.: Uniform expression of stress-strain relationship of soils. J. Eng. Geol. **18**(6), 900–905 (2010)

Wang, L., Zhao, Z., Li, L.: Modified Dunean-Chang model considering soil strueture. J. Hydraul. Eng. **1**, 83–89 (2004)

Chen, Z.: Principle, Method and Procedure of Soil Slope Stability Analysis. China Water Conservancy and Hydropower Press, Beijing (2003)

Hydro-Thermal Properties of the Unsaturated Soil

Fujiao Tang and Hossein Nowamooz(✉)

INSA de Strasbourg, 24 Boulevard de la Victoire, 67084 Strasbourg Cedex, France
Hossein.Nowamooz@insa-strasbourg.fr

Abstract. In this paper, 10 approaches were initially used to predict the thermal conductivity (k) of different soils. The comparison showed that three principal parameters indicating sand content (x_s), dry density (ρ_{dry}), and degree of saturation (S_r) influenced highly the soil thermal conductivity. Moreover, 3 approaches for the volumetric heat capacity (C_v) of soil were used to predict the experimental data from the literature. The result showed that the classical approaches can induce the errors because of the non-consideration of the mineral and water content. This insufficiency was solved by proposing a new model. Finally, the most compatible approaches for the thermal properties were implemented into a 2D subsurface model using finite element method (FEM). The variation of suction (s), thermal conductivity (k) and temperature (T) with time and space was then investigated in the numerical simulation model under the influence of seasonal suction and temperature on the top boundary.

1 Introduction

The heat transport process in porous media is governed by thermal properties, i.e. thermal conductivity, heat capacity, and thermal diffusivity (Saito et al. 2014). The thermal properties in soil have a significant influence on the agriculture, biology, soil mechanics, meteorology and new energies. To be specific, the soil thermal properties have received attention for more than one century (Forbes 1849). At first, the soil thermal properties are mainly concentrated on the mechanical and agricultural behavior of soil. In 1912, the first known record of the concept of using the ground as a heat source for a heat pump is found in a Swiss patent (Ball et al. 1983). During those years, due to the lack of geothermal energy usage, the soil thermal properties did not attract the public attention. It is not until the 1970s after the first oil crisis that geothermal energy has drawn up more scientists' eyes (Yang et al. 2010). From the late of the 20th century till now, the geothermal engineering developed faster than ever and there is a significant development of the heat transfer in soils.

As the heat transfer in soils can be influenced by its mineral composition, water content, dry density, temperature and pressure. It is necessary to study the thermal underground properties to improve shallow borehole heat exchanger (SBHE) performance. Thermal response test (TRT) is now widely used all around the world in geothermal engineering. Using TRT in the site can help engineers to have a general understanding of the thermal properties. However, it cannot get the variation of

S. Wang et al. (eds.), *New Solutions for Challenges in Applications of New Materials and Geotechnical Issues*, Sustainable Civil Infrastructures,
https://doi.org/10.1007/978-3-319-95744-9_2

hydro-thermal properties with time and space. As well, TRT result is merely a result for a given time and space and money consuming. Therefore, appropriate prediction models for hydro-thermal properties are welcomed in this field.

2 Proposed Approaches for the Thermal Properties

In this paper, we assumed that the soil thermal properties were influenced by the sand content, the degree of saturation and the dry density. A least square method (R^2) was also employed to illustrate the capacity of the approaches.

2.1 Soil Thermal Conductivity Approach

All the exiting approaches for the soil thermal conductivity were outlined in Table 1.

The performance of 10 approaches was compared with experimental data from the literature for 7 different soils: sand (Alrtimi et al. 2016; Abu-Hamdeh and Reeder 2000; Lu et al. 2007; Barry-Macaulay et al. 2013), loam (Lu et al. 2007), clay loam (Abu-Hamdeh 2001; Lu et al. 2007; Barry-Macaulay et al. 2013), silt loam (Lu et al. 2007; Barry-Macaulay et al. 2013), bentonites (Tang et al. 2008), sandy clay loam (Barry-Macaulay et al. 2013), and silt clay loam (Barry-Macaulay et al. 2013). 4 approaches with the highest correlation coefficient were shown in Fig. 1. In addition, the correlation coefficients (R^2) of 10 models were reported in Fig. 1 to facilitate the comparison. The results showed the approach proposed by Nowamooz et al. (2015) had the highest correlation coefficient among all the aforementioned models. Furthermore, it needed only three parameters.

2.2 Volumetric Heat Capacity Approach

Soil volumetric heat capacity is the coupling of minerals, water, air and vapor. Table 2 summarized the equations and parameters of two existing approaches as well as a newly proposed approach.

Figure 2 showed the estimations of 3 approaches compared with the experimental data reported by Abu-Hamdeh, 2003 (sand and clay). The results showed that the existing approaches were not the best choice due to the non-consideration of the mineral content and water content. This insufficiency was solved by proposing a new approach for the volumetric heat capacity (C_v) with a better correlation coefficient.

3 Characterization of the Hydro-Thermal Soil Behavior

Studying heterogeneous soil could have the benefits of distinguishing the different soil hydro-thermal properties and evaluating the capacity of the proposed models for the multi-layered soils. A 2D finite element method code (COMSOL) was used to study the variation of the temperature (T) and soil suction (s) at different depths of a multi-layered soil with time.

Table 1. Approaches proposed for the thermal conductivity

Approach	Equation	Parameter[a]
Kersten (1949)	$k = \left\{(a_1 \log \omega - b_1) \times 10^{0.6242\rho_d - 3.4628}\right\} \times 418.6$	a_1, b_1, w, ρ_d
de Vries (1963)	$x_{a\text{-}v} \leq 0.5$: $k = \frac{x_{w-v}k_w + F_a x_{a-v}k_a + F_p x_{p-v}k_p}{x_{w-v} + F_a x_{a-v} + F_p x_{p-v}}$ $x_a > 0.5$: $k = \frac{x_a k_a + F_{pw}(x_w + x_p)k_{pw}}{x_a + F_{pw}(x_w + x_p)}$	$x_{a\text{-}v}$, $x_{w\text{-}v}$, $x_{p\text{-}v}$, k_a, k_w, k_p, $k_{p\text{-}w}$, F_a, F_p, F_{pw}
Johansen (1975)	$k_p = k_q^{x_{q-v}} k_o^{1-x_{q-v}}$ $k_{sat} = k_p^{(1-n)} k_w^n$ $k_{dry} = \frac{0.135\rho_d + 64.7}{2700 - 0.947\rho_d} \pm 20\%(soils)$ $k_{dry} = 0.039n^{-2.2} \pm 25\%(rocks)$ $k = (k_{sat} - k_{dry}) \times (m \log S_r + 1.0) + k_{dry}$	k_p, k_q, $x_{q\text{-}v}$, m, k_o, k_w, n, S_r
Sakashita and Kumada (1998)	$k_{dry} = 0.0479 + 0.222(1 - n) + 0.968(1 - n)^3$ $k = k_{dry}\{1 + [(9.750n - 0.706)S_r]^{0.285n+0.731}\}$	n, S_r
Coté and Konrad (2005)	$k = (k_p^{(1-n)} k_w^n - \chi \times 10^{-\eta n})k_e = \left[\frac{aS_r}{1+(a-1)S_r}\right] + \chi \times 10^{-\eta n}$	k_p, χ, η, n, k_w, k_p, a, S_r
Balland and Arp (2005)	$k_{sat} = k_p^{(1-n)} k_w^n$ $k_{dry} = \frac{(ak_p - k_a)\rho_b + k_a\rho_p}{\rho_p - (1 - 0.053)\rho_d}$ $k_e = S_r^{0.5(1 + x_{om-v} - 0.24x_{s-v} - x_{cf-v})}\left[\left(\frac{1}{1 + \exp(-18.3S_r)}\right)^3 - \left(\frac{1-S_r}{2}\right)^3\right]^{1-x_{om}}$ $k = (k_{sat} - k_{dry})k_e + k_{dry}$	k_p, n, k_w, k_a, ρ_b, ρ_p, S_r, $x_{om\text{-}v}$, $x_{s\text{-}v}$, $x_{cf\text{-}v}$
Lu et al. (2007)	$k_p = k_q^{x_{q-v}} k_o^{1-x_{q-v}}$ $k = [k_p^{1-n} k_w^n - (0.51 - 0.56n)] \exp[\alpha(1 - S_r^{\alpha-1.33}]$ $+ 0.51 - 0.56n$	k_q, k_o, $x_{q\text{-}v}$, n, α, S_r
Chen (2008)	$k(n, S_r) = k_p^{1-n} k_w^n[(1 - b)S_r + 0.0022]^{0.78n}$	k_p, n, k_w, S_r
Haigh (2012)	$a_a = k_a/k_p$ $a_w = k_w/k_p$ $\xi = \frac{2n-1}{3}$ $\frac{k}{k_p} = 2(1+\xi)^2 \left\{ \begin{array}{l} \frac{a_w}{(1-a_w)^2} \ln\left[\frac{(1+\xi)+(a_w-1)x}{\xi + a_w}\right] + \\ \frac{a_a}{(1-a_a)^2} \ln\left[\frac{(1+\xi)}{(1+\xi)+(a_a-1)x}\right] \end{array} \right\}$ $+ \frac{2(1+\xi)}{(1-a_w)(1-a_a)}[(a_w - a_a)x - (1 - a_a)a_w]$	k_a, k_w, k_p, n
Nowamooz et al. (2015)	$k = (0.443x_s + 0.081\gamma_d)\frac{(4.4x_s + 0.4)S_r}{1 + (4.4x_s - 0.6)S_r} + 0.087x_s + 0.019\gamma_d$	γ_d, x_s, S_r

[a] a_1, b_1, m, α, x, F_a, F_p, F_{pw} are shape factors; w is gravimetric water content; ρ_d is dry density (g/cm^3); $x_{a\text{-}v}$, $x_{w\text{-}v}$, $x_{p\text{-}v}$, $x_{om\text{-}v}$, $x_{s\text{-}v}$, $x_{cf\text{-}v}$, $x_{j\text{-}v}$ are volumetric content of air, water, porous particle, organic matter, sand, coarse fragment and different materials; k_{pw} is weighing factor; k_w, k_a, k_p, k_q, k_o, k_{mj} are thermal conductivity of water, air, soil particle, quartz, other minerals and different materials (w m^{-1} k^{-1}); k_{sat} and k_{dry} are saturated and dry thermal conductivity (w m^{-1} k^{-1}); S_r is soil saturation; n is soil porosity; χ, η and a are parameters related to soil mineral content; x_s is gravimetric sand content; γ_d is soil dry density multiplied by gravity (kN m^{-3})

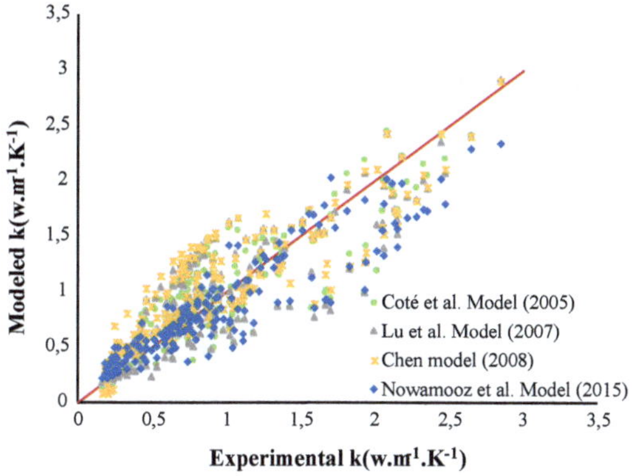

Model	R^2	Model	R^2
Kersten	0.640	Balland	0.366
de Vries	0.618	Lu	0. 669
Johansen	0.516	Chen	0.663
Sakashita	0.074	Haigh	0.074
Coté	0.744	Nowamooz	0.762

Fig. 1. Comparison between the different approaches proposed for the thermal conductivity

Table 2. Approaches proposed for the soil volumetric heat capacity

Approach	Equation	Parameter[1]
de Vries et al. (1963)	$C_v = 0.46x_{p-v} + 0.60x_{om-v} + x_{w-v}$ (cal/cm^3 °C)	x_{p-v}, x_{om-v}, x_{w-v}
Ghuman et al. (1985)	$C_v = \rho_b e^{-0.003X-1.071}$ (cal/(g °C))	ρ_b, X
Newly developed approach	$C_v = (4.18 - 0.95\rho_d - 0.3x_s)$ $S_r + 0.9\rho_d - 0.2x_s$	ρ_d, x_s, S_r

[1] x_{p-v}, x_{om-v}, x_{w-v} are volumetric water content of particle, organic matter, and water; ρ_b and ρ_d are soil bulk density and soil dry density (g cm^{-3}); X is the sum of sand, silt and organic matter in percentages; x_s is sand content by mass; S_r is soil saturation

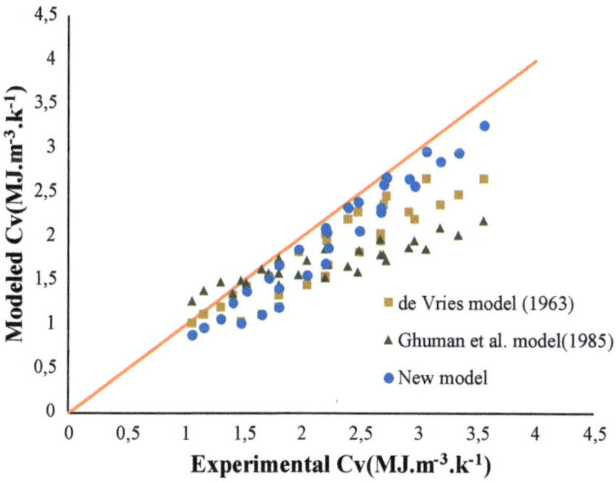

Model	R^2
De Vries	0.4776
Ghuman	-0.0979
Newly developed	0.8016

Fig. 2. Comparison between different approaches proposed for the volumetric heat capacity

3.1 Model Description

The 2D geometry was made up of loamy sand and loam (Fig. 3). The generated mesh contained 55678 triangular elements.

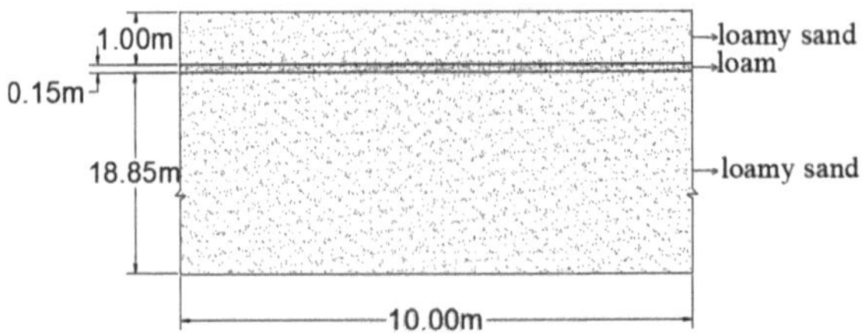

Fig. 3. Model geometry

Richards Eq. (1931) was used for calculating the variation of water potential with time and space, which is based on the conservation of mass. It can be expressed as:

$$C\frac{\partial H_p}{\partial t} + \nabla \cdot \left[-K \cdot k_r \cdot \nabla\left(H_p + D\right)\right] = 0 \tag{1}$$

where, C is specific moisture capacity (1/m), H_p is water potential or suction (m), K is hydraulic conductivity (m/s) of porous media, D is elevation head (m). Mualem Eq. (1976) was used for the calculation of relative hydraulic conductivity k_r:

$$k_r = \begin{cases} S_e^l\left[1 - \left(1 - S_e^{n/(n-1)}\right)^{1-1/n}\right]^2 & H_p < 0 \\ 1 & H_p \le 0 \end{cases} \tag{2}$$

where, relative saturation S_e is in the following form:

$$S_e = \frac{\theta - \theta_r}{\theta_s - \theta_r} \tag{3}$$

where, θ, θ_s and θ_r are volumetric water content, saturated volumetric water content and residual volumetric water content, respectively; n is van Genucheten parameter; l is pore conductivity parameter.

van Genucheten (1980) equation was used for the calculation of suction:

$$S_e = \begin{cases} \frac{1}{\left[1 + \left|\alpha H_p\right|^n\right]^{1-1/n}} & H_p < 0 \\ 1 & H_p \ge 0 \end{cases} \tag{4}$$

where, H_p is water potential or suction (m); α (m^{-1}) is van Genucheten parameter.

Fourier's equation was in parallel used for heat transfer in soil:

$$C_v \frac{\partial T}{\partial t} = \nabla \cdot (k \nabla T) \tag{5}$$

where, t is time. It is worth noting that volumetric heat capacity (C_v) and thermal conductivity (k) varied with hydraulic conditions described in the previous section.

All the parameters of the numerical modeling were summarized in Table 3, where ρ_p denotes the particle density (g/cm^3).

Table 3. Input material properties of the numerical model

Material	ρ_{dry}	χ_s	K	l	α	n	θ_s	θ_r	ρ_p
Loamy sand	1.63	0.8	1e − 5	0.5	5	1.5	0.39	0.04	2.65
Loam	1.43	0.5	1e − 8	0.5	1.3	1.1	0.46	0.06	2.65

We assumed that the initial soil was saturated and the underground temperature was initially 10 °C. On the top boundary, the suction was set to change with time. The surface suction in winter was the lowest, which reached to 0 MPa. Meanwhile, the soil surface suction in summer was the highest, which reached 2 MPa (Fig. 4a). In addition, the top boundary was also influenced by the seasonal temperature variation during one year as presented in Fig. 4b. It was supposed that there was neither water nor temperature flow on the other boundaries.

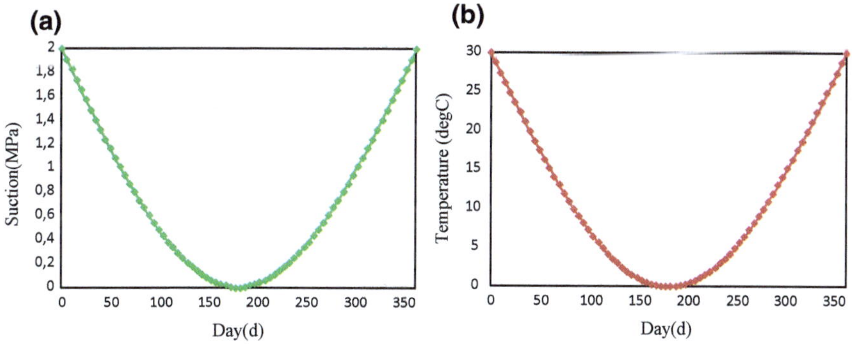

Fig. 4. **a** Suction (MPa) and **b** temperature (°C) applied on the soil surface

3.2 Simulation Results

By implementing the best approaches for the thermal conductivity and thermal capacity (models presented in Tables 1 and 2 with the highest correlation in Figs. 1 and 2) into the finite element numerical model, the simulation results were derived with time and space. The variation of the suction (s), the thermal conductivity (k) and the temperature

(T) was investigated with depth after 90 days in Fig. 5a, 182 days in Fig. 5b and 270 days in Fig. 5c. It should be mentioned that the soil volumetric heat capacity was proportional to the saturation in the newly-developed model and the soil thermal properties affected the soil temperature. It could be concluded from Fig. 5:

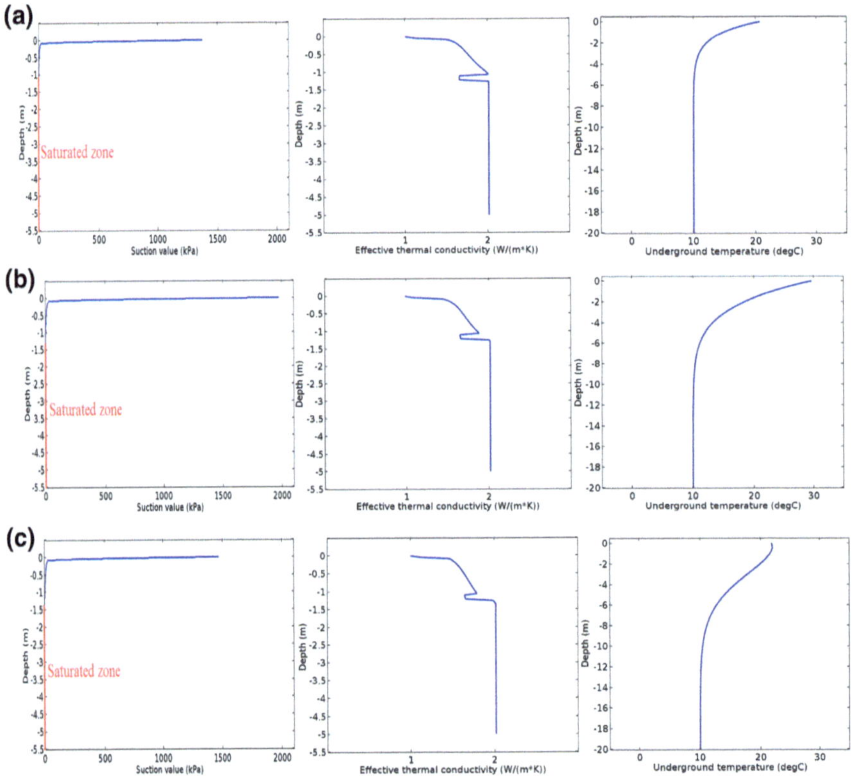

Fig. 5. Suction, thermal conductivity and temperature at different times: **a** t = 90 day; **b** t = 182 day; **c** t = 270 day

(1) The suction fluctuation on the land surface of the model was in a depth between 0 and 1.5 m.
(2) The soil thermal conductivity was related to the soil hydraulic condition (water content). Moreover, it was dependent on its sand content and dry density.
(3) In these simulations, the layer of sandy loam had a higher thermal conductivity than the layer of loamy soil with the same water content due to its higher dry density and sand content.
(4) The soil temperature changed with time and space. For a higher depth, the soil temperature was less dependent on the surface temperature.

4 Conclusions

By comparing 10 models for the soil thermal conductivity, it was concluded that the model proposed by Nowamooz et al. (2015) was the most convenient. By utilizing literature results, a new model for the soil volumetric heat capacity was additionally developed based on the soil degree of saturation (S_r), the particle dry density (ρ_{dry}) and the sand content (χ_s). The new model showed a better performance compared to the existing models.

A 2D model considering the seasonal hydro-thermal solicitations was then built. It could be seen from the simulation results that the hydro-thermal properties changed with time and space, and the selected approaches for the thermal conductivity and capacity were capable to model the hydro-thermal transfer in soil and to predict the temperature variations with time and space.

Acknowledgments. This work is supported by the China Scholarship Council (CSC).

References

Abu-Hamdeh, N.H.: Measurement of the thermal conductivity of sandy loam and clay loam soils using single and dual probes. J. Agric. Eng. Res. **80**(2), 209–216 (2001)

Abu-Hamdeh, N.H., Reeder, R.C.: Soil thermal conductivity: effects of density, moisture, salt concentration, and organic matter. Soil Sci. Soc. Am. J. **64**, 1285–1290 (2000)

Alrtimi, A., Rouainia, M., Haigh, S.: Thermal conductivity of a sandy soil. Appl. Therm. Eng. **106**, 551–560 (2016)

Ball, D.A., Fischer, R.D., Hodgett, D.L.: Design methods for ground-source heat pumps. ASHRAE Trans. **89**(2B), 416–440 (1983)

Balland, V., Arp, P.: Modelling soil thermal conductivities over a wide range of conditions. J. Environ. Eng. Sci. **4**, 549–558 (2005)

Barry-Macaulay, D., Bouazza, A., Singh, R.M., Wang, B., Ranjith, P.G.: Thermal conductivity of soils and rocks from the Melbourne (Australia) region. Eng. Geol. **164**, 131–138 (2013)

Chen, S.: Thermal conductivity of sands. Heat Mass Transf. **44**(10), 1241–1246 (2008)

Coté, J., Konrad, J.M.: A generalized thermal conductivity model for soils and construction materials. Canad. Geotech. J. **42**(2), 443–458 (2005)

de Vries, D.A.: Thermal properties of soil. In: van Wijk, W.R. (ed.) Physics of Plant Environment, pp. 210–235. New Holland, Amsterdam (1963)

Forbes, J.D.: Account of some experiments on the temperature of the Earth at different depths, and in different soils, near Edinburgh. Trans. Roy. Soc. Edinb. **16**, 189–236 (1849)

Haigh, S.K.: Thermal conductivity of sands. Geotechnique **62**(7), 617–625 (2012)

Johansen, O.: Thermal conductivity of soils, PhD Thesis, Trondheim, Norway, ADA 044002 (1975)

Kersten, M.: Thermal properties of soils. Eng. Exp. Stat. **52** (1949)

Lu, S., Ren, T., Gong, Y., Horton, R.: An improved model for predicting soil thermal conductivity from water content at room temperature. Soil Sci. Soc. Am. J. **71**(1), 8–14 (2007)

Mualem, Y.: A new model for predicting the hydraulic conductivity of unsaturated porous media. Water Resour. Res. **12**(3), 513–522 (1976)

Nowamooz, H., Nikoosokhan, S., Jian Lin, B., Chazallon, C.: Finite difference modeling of heat distribution in multilayer soils with time-spatial hydrothermal properties. Renew. Energy **76**, 7–15 (2015)

Richards, L.A.: Capillary conduction of liquids through porous mediums. J. Appl. Phys. **1**, 318–333 (1931)

Saito, T., Hamamoto, S., Mon, E.E., Takemura, T., Saito, H., Moldrup, T.K.P.: Thermal properties of boring core samples from the Kanto area, Japan: development of predictive models for thermal conductivity and diffusivity. Soils Found. **54**(2), 116–125 (2014)

Sakashita, H., Kumada, T.: Heat transfer model for predicting thermal conductivity of highly compacted bentonite. J. Japan At. Soc. **40**, 235–240 (1998)

Tang, A.M., Cui, Y.-J., Le, T.T.: A study on the thermal conductivity of compacted bentonites. **41**(3–4), 181–189 (2008)

van Genuchten, M.T.: A closed-form equation for predicting the hydraulic conductivity of unsaturated soils. Soil Sci. Soc. Am. J. **44**, 892–898 (1980)

Yang, H., Cui, P., Fang, Z.: Vertical-borehole ground-coupled heat pumps: a review of models and systems. Appl. Energy **87**, 16–27 (2010)

Gas Bubble Nucleation and Migration in Soils—Pore-Network Model Simulation

Nariman Mahabadi[1], Xianglei Zheng[2], Tae Sup Yun[3], and Jaewon Jang[4(✉)]

[1] Arizona State University, Tempe, AZ, USA
[2] Municipal Testing Laboratory, New York, USA
[3] Yonsei University, Seoul, South Korea
[4] Hanyang University, Seoul, South Korea
jwj@hanyang.ac.kr

Abstract. Sediment can be de-saturated by introducing gas bubbles, which is found in various applications such as methane gas generation in landfill, microbial-induced gas bubble formation, air sparing method for soil remediation, heavy oil depressurization for carbon recovery, and gas production from hydrate bearing sediment. The gas introduction method (e.g., nucleation and injection) and migration and trapping of gas bubbles affect the hydraulic conductivity, residual gas saturation, and the stability of these gassy sediments. In this study, the pore-network model is used to investigate gas bubble migration in porous media. Gas bubbles are introduced by mimicking either nucleation or injection. Based on the known gas bubble behavior available in the literature, numerical algorithms are developed to simulate the migration and trapping of gas bubbles in pore-network model. The effect of gas bubble size distribution and pore size distribution on residual saturation is investigated. The results show that gas bubble size distribution becomes wider as gas bubbles coalesce to each other during migration. And the residual gas saturation increase with increasing bubble size and permeability reduction becomes apparent as the gas bubble size and the number of generated gas bubble increase.

1 Introduction

Soils can be de-saturated by several gas formation mechanisms such as microbial activity in shallow ocean sediments or wetlands, methanogenic degradation of hydrocarbon contaminants in the subsurface (Amos et al. 2005), decomposition of municipal solid waste in landfills (van Breukelen et al. 2003), air trapping by groundwater-level oscillation (Krol et al. 2011), and seasonal temperature variation resulting in gas solubility change in the subsurface (Ryan et al. 2000). In addition, there is a possibility of gaseous CO_2 formation by the leakage-induced depressurization of CO_2-dissolved brine during the long-term geological CO_2 sequestration (Plampin et al. 2014; Zuo et al. 2012; Zuo et al. 2013).

On the other hand, gas bubbles can be also introduced artificially to remediate contaminated soils, modify the properties of the sediments, and produce resources in various applications such as air sparging or gas exsolution by supersaturated water

S. Wang et al. (eds.), *New Solutions for Challenges in Applications of New Materials and Geotechnical Issues*, Sustainable Civil Infrastructures,
https://doi.org/10.1007/978-3-319-95744-9_3

injection (SWI) for soil remediation (Enouy et al. 2011; McCray and Falta 1997), denitrification and induced partial saturation (IPS) for liquefaction prevention (Eseller-Bayat et al. 2013; He and Chu 2014; Rebata-Landa and Santamarina 2012), heavy oil depressurization to reduce viscosity (e.g., solution gas drive) (Bora et al. 2000; Stewart et al. 1954), methane gas production from hydrate-bearing sediments (Jang and Santamarina 2011, 2014; Jang and Santamarina 2016), and CO_2 sequestration/CO_2 foam injection (Zheng and Jang 2016; Zheng et al. 2017). The gas generation mechanisms in the abovementioned applications include direct gas bubble injection, depressurization, temperature increase, electrolysis, and drainage-recharge.

Once the gas bubbles are generated in the sediment, they can migrate upward due to the buoyancy, or are sometimes trapped in the pore space. The gas nucleation, migration, and trapping and the associated effects are frequently found in the in situ sediment. Methane ebullition, the release of methane into the atmosphere or the movement through porous media, is the typical mechanism of greenhouse gas emission from aquatic ecosystems (Amos and Mayer 2006; Ramirez et al. 2015; Walter et al. 2006). Sometimes, methane bubbles burst out and form a crater in the permafrost gradually thawing due to the global warming (Moskvitch 2014). The gas bubble formation in the shallow ocean sediment also affects the mechanical properties of the sediment (Grozic et al. 1999; Sills et al. 1991). In addition, very small gas bubbles trapped in the porous media can dramatically reduce hydraulic conductivity without the significant reduction in water saturation (Ronen et al. 1989).

The initial size of gas bubbles upon nucleation, the coalescence of gas bubbles during migration, the bubble generation rate, and the pore throat size of the sediment could affect the behavior of gas migration and trapping in the porous media. In this study, we studied the behavior of gas bubble migration in the porous media and investigated the effect of gas bubble size on the residual gas saturation and hydraulic conductivity.

2 Simulation Details

The migration of gas bubbles through the porous media is simulated using the pore network model extracted from 3D X-ray CT image of soils. Several numerical algorithms and criteria for the size-dependent velocity of rising gas bubbles, bubble coalescence, escaping, and trapping in the pore space are summarized in this section.

2.1 Pore Network Model Extraction from X-Ray CT Image

A sediment core used for the X-ray scanning was recovered from Mallik 5L-38 site in Beaufort Sea, Canada (The grain size distribution of the sediment and the information on the X-ray scanning is available in Mahabadi et al. (2016a) and Mahabadi et al. (2016b). The volume of the scanned image is 27 mm^3 (3 mm × 3 mm × 3 mm) with 12.5 μm pixel resolution. The obtained CT images provide the three-dimensional structure of the scanned sediment, including both the grains and the pore spaces (Fig. 1a). Then, the maximal ball algorithm developed by Dong and Blunt (2009) is employed to extract the three-dimensional pore-network model from the X-ray CT images (Mahabadi and Jang 2014; Mahabadi et al. 2016b) (Fig. 1a). The maximal ball algorithm generates spheres inscribed in the pore wall. Then, the bigger spheres turn into the pores of the pore-network model and the size of smaller spheres located in pore throats are used as the radii of cylindrical tubes connecting the neighboring pores. As a result, the pore-network model consists of the spherical pores connected by cylindrical tubes. The extracted pore network model consists of 4593 pores and 19361 tubes with the tube connectivity per pore (coordination number) of cn = 8.0. Figure 1b shows the pore and tube size distribution of the extracted pore network model.

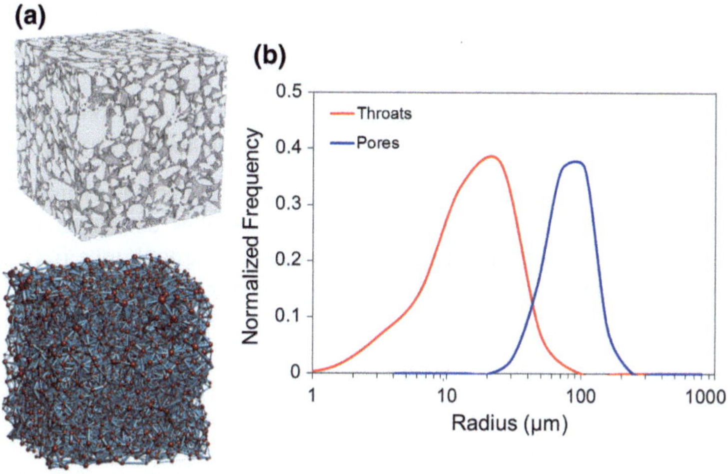

Fig. 1. Details of pore-network model simulation. (a) CT-Scan images from 3 mm × 3 mm 3 mm sample of Mallik Sand, and pore network model extracted from the CT-scan image. The extracted pore network model consists of 4593 pores and 19361 tubes with the tube connectivity per pore (coordination number) of cn = 8.0. Mean pore radius is $\mu[R_{pore}] = 69.3$ μm, mean tube size is $\mu[R_{tube}] = 12.5$ μm, and mean tube length $\mu[L_{tube}] = 45$ μm ($\max[R_{pore}] = 194$ μm, $\min[R_{pore}] = 22$ μm, $\max[R_{tube}] = 63$ μm, $\min[R_{tube}] = 1$ μm). (b) Pore and throat size distribution.

2.2 Bubble Generation and Migration

Bubble generation. Gas bubbles are nucleated at randomly selected pores in the pore-network model. Only one gas bubble is generated per a pore. The ratios of the number of the nucleated gas bubbles N_B over the total number of pores N_P used in this simulation include $N_B/N_P = 0.2$, 0.4, 0.6, and 0.8. Regarding the bubble size, either (1) mono-sized bubbles are nucleated at the selected pores or (2) the sizes of bubbles are determined by the size of pores that occupy the bubbles. The size of mono-sized bubbles used in the simulation ranges from $R_B = 5$ μm to 80μm. And the ratio of the bubble radius R_B over the host pore radius R_P used in this study ranges from $R_B/R_P = 0.125$ to 0.5.

Rising bubble velocity. The velocity of a spherical solid particle settling in water is derived from the Stokes' law (ASTM 2016) in which the gravitational force of the particle is assumed to be the same as the viscous drag force by the fluid, resulting in the terminal velocity V_s [m/s]:

$$V_s = \frac{g(\rho_s - \rho_w)d^2}{18\mu} \tag{1}$$

where $g[m/s^2]$ is the gravitational acceleration, $\rho_s[kg/m^3]$ is the particle density, $\rho_w[kg/m^3]$ is the water density, d [m] is the particle diameter, and $\mu[kg/(m\ s)]$ is the dynamic viscosity of the fluid. For the case of a gas bubble ascending in a fluid due to the buoyancy, the velocity of the rising bubble V_B in water can be estimated from Eq. 1 by assuming the mass density of the gas bubble to be zero.

As a gas bubble moves upward, it is expected that the bubble size increases due to the decreasing hydrostatic pressure. But, in this study, the size of the gas bubble is assumed to be constant due to the small change in hydrostatic pressure within a 3 mm in height of the pore network model [Note that the size of the air bubble increases 3%–5% while it moves along 1 m in vertical distance (Roosevelt and Corapcioglu 1998)]. Therefore, once a bubble nucleates, a constant velocity is assigned to the bubble.

Bubble migration, coalescence, and trapping. Once a gas bubble is assigned in a pore, the gas bubble migrates upward through one of the tubes connected to the pore. Consider a pore i P_i connected to neighboring pores j P_j through tube ij T_{ij}. For each pore j connected to the pore i, the following value is calculated:

$$\sin\theta = \frac{(z_j - z_i)}{L_{ij}} \tag{2}$$

where z_j and z_i are z-coordinates of P_i and P_j, and L_{ij} is the length of the tube ij T_{ij} connecting P_i and P_j. Therefore, sinθ means the vertical gradient of the tube along z-axis ($-1 \leq \sin\theta \leq 1$). All tubes connected to P_i are ranked based on sinθ values. It is assumed that the bubble migrates through the tube that has the highest sinθ value. If the tube has a negative sinθ value, the bubble will not migrate through the tube due to the buoyancy. Based on this criterion, the tube with highest sinθ-value is selected as the pathway for bubble migration unless the P_j or T_{ij} is blocked by a gas bubble. When the radius of the bubble inside P_i is greater than the radii of all T_{ij} connected to P_i, the gas bubble is considered as trapped in the pore. If there is a trapped gas bubble in a pore, another bubble could migrate into the pore and coalesces to the existing trapped gas bubble as long as the coalesced gas bubble size is smaller than the pore size.

A time step Δt_m is selected such that only one gas bubble (moving from P_i to P_j along the tube ij T_{ij}) is allowed to arrive at the neighboring pore j during the time step:

$$\Delta t_m = \min\left[\frac{\Delta L_{ij}}{v_{ij}}\right] \tag{3}$$

where v_{ij} is the velocity of the rising bubble in the tube ij T_{ij} connecting P_i and P_j and ΔL_{ij} is the distance from the gas bubble in the T_{ij} to the neighboring pore j P_j. Therefore, only one bubble can reach to the neighboring pore at each time step unless there is a coalescence between two bubbles in a tube.

The coalescence of gas bubbles can happen in the tube. Consider two gas bubbles, B_1 and B_2, migrating in the same tube. At the time $t = t_1$, the locations of small bubble

B_1 and large bubble B_2 are L_1 and L_2 from the input side (bottom) of the tube, respectively. The coalescence of the two bubbles occurs at the time $t = t_1 + \Delta t_c$ for the condition below,

$$(L_1 + V_1 \Delta t_c) - (L_2 + V_2 \Delta\, t_c) = R_{B1} + R_{B2} \tag{4}$$

where V_1 and V_2 are the velocity of the rising bubbles B_1 and B_2 which are dependent on the their sizes, and R_{B1} and R_{B2} are the radii of the bubble B_1 and B_2.

The time step for bubble coalescence Δt_c is calculated from Eq. 4:

$$\Delta\, t_c = \frac{L_1 - (L_2 + R_{B1} + R_{B2})}{V_2 - V_1} \tag{5}$$

The minimum coalescence time step Δt_c is chosen such that only one coalescence event is allowed to occur for all the bubbles moving in all the tubes. When the two bubbles are merged together and form a bigger bubble, the velocity of the merged bubble is calculated based on its new size. If the radius of the merged bubble is bigger than the radius of the tube, the gas bubble is considered to be trapped in the tube.

Once Δt_c is calculated, the smaller time step between Δt_m and Δt_c is chosen for the global time step $\Delta t = \min[\Delta t_m, \Delta t_c]$. Based on the calculated time step Δt, the location of the bubbles in the pore-network model is updated during the migration and this procedure is repeated until there is no further movement of gas bubbles in the pore-network model. During the migration, the bubbles arriving at the outlet pores in the top layer escape from the pore network model.

Hydraulic conductivity. Once a gas bubble is trapped in a pore, the tubes connected to the pore loose conductivity. Therefore, for the conductivity calculation, the pore and the neighboring tubes are removed from the pore network model assuming zero conductivity. At the end of each gas migration simulation, the hydraulic conductivity is calculated (Jang et al. 2011) and normalized by the hydraulic conductivity obtained for 100% water saturation condition.

3 Results and Analyses

In this study, the effect of bubble size on the gas bubble migration in the porous media is investigated. Two cases of the bubble size distribution are considered: (1) mono-sized bubbles, and (2) distributed-sized bubbles. The size of bubbles are varied from $R_B = 5$ μm to 65 μm for the mono-size bubble case. And for the distributed-size bubble case, the ratio of the bubble radius R_B over the host pore radius R_P varies from $R_B/R_P = 0.125$ to 0.5.

Gas bubbles are initially assigned to the randomly chosen pores such that each pore occupies only one bubble. The ratios of the number of pores that occupy gas bubbles over the total number of pores in the pore-network model are $N_B/N_P = 0.2$, 0.4, 0.6 and 0.8. The generated gas bubbles start migrating upward towards the outlet in the top layer due to the buoyancy. Some of gas bubbles are trapped in the pore-network model, which determines the final gas saturation after the simulation. Initial and final images during gas bubble migration for mono-sized bubble case ($R_B = 15$ μm, $N_B/N_P = 0.4$) are shown in Fig. 2. During the bubble migration, a rising bubble can be merged with the bubble already trapped in the upper pore and form a bigger bubble.

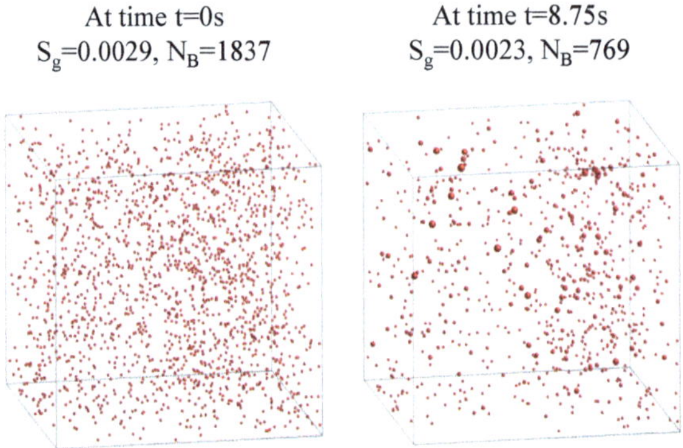

Fig. 2. Gas bubble location during gas migration and coalescence. Mono-size bubbles for bubble size $R_B = 15$ μm and $N_B/N_P = 0.4$ ($N_B = 1837$, $N_P = 4593$). Left column shows the initial nucleation of bubbles and the right column shows the final gas bubble saturation.

During gas bubble migration, some gas bubbles escape from the pore-network model when they reach to the top layer, some gas bubbles are trapped inside the pore-network, and some gas bubbles are coalesced to each other forming bigger gas bubbles. Therefore, the gas bubble size distribution changes and the number of total gas bubbles N_B decrease (Fig. 3). For mono-sized case, initially, there were a total of 1787 uniform-sized ($R_B = 15$ μm) gas bubbles. And there occur bigger gas bubbles during the simulation due to the coalescence and the size distribution curve becomes wider. The total number of gas bubble at the end of the simulation is $N_B = 769$. And for distributed-size case, the number of gas bubbles smaller than $R_B \sim 25$ μm decreases (possibly due to escaping and coalescence) and the number of gas bubbles larger than $R_B \sim 25$μm increases during the simulation.

The total migration time increases as the size of gas bubbles decreases regardless of the initial number of bubbles for mono-size bubble case study. In this case, all the bubbles move at the same ascending velocity as long as their sizes are the same (predicted by Eq. 1). As a result, the total migration time is only a function of the initial bubble size for a given pore network model dimension. However, the total migration time for distributed-sized bubble case also increases as R_B/R_P decreases. For distributed sized bubble case, the larger number of nucleated gas bubbles (higher N_B/N_P) results in wider bubble size distribution which means $N_B/N_P = 0.8$ case could include some bubbles smaller than the smallest bubble and larger than the largest bubble generated for the $N_B/N_P = 0.2$ case. For the low R_B/R_P ratio (e.g. $R_B/R_P = 1/8$), the smaller gas bubbles that have low ascending velocity require longer time to escape from the pore-network model. For the higher R_B/R_P ratio (e.g. $R_B/R_P > 1/5$), the large bubbles tend to block the pores and prevent the migration of bubbles, which reduces in total migration time.

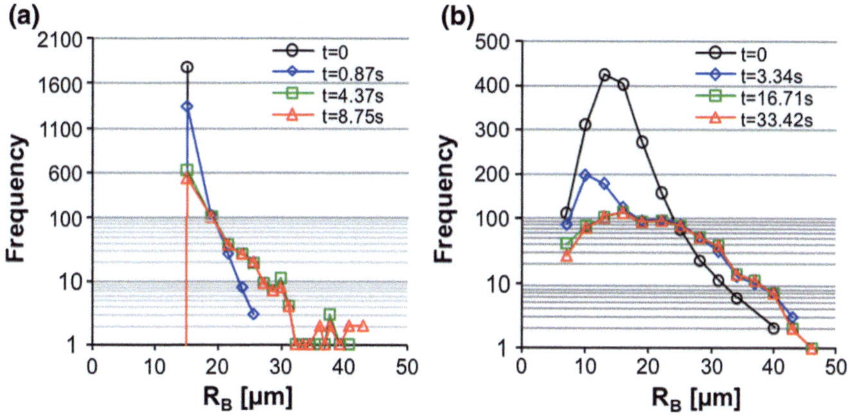

Fig. 3. Statistical and spatial bubble size distribution during gas bubble migration for the case of $N_B/N_P = 0.4$. (a) Mono-sized case (the initial size of gas bubble $R_B = 15$ μm). (b) Distributed-sized case (the ratio of pore size over bubble size $R_B/R_P = 1/5$). The change in gas bubble size distribution is shown during the migration due to coalescence.

If the radius of the bubble nucleated in a certain pore is bigger than the radii of any other tubes connected to the pore, the bubble is trapped in the pore. If another gas bubble migrates into the pore that occupies the trapped bubble, two gas bubbles are merged together and form a bigger trapped bubble as long as the volume of the coalesced gas bubble is smaller than the pore volume.

The residual gas saturations, the volume of trapped gas bubbles divided by the total volume of pore space, for mono-sized and distributed-size bubble cases are shown in first row in Fig. 4. The higher number of initial bubbles (higher N_B/N_P ratio) results in the higher residual gas saturation for a given bubble size.

When a gas bubble is trapped in a pore, the tubes connected to the pore loose the conductivity. Therefore, as more gas bubbles are trapped in the pore-network model, the global hydraulic conductivity of the pore-network model decreases. The hydraulic conductivity of the pore-network model with the trapped gas bubbles at the end of the simulation is normalized by the hydraulic conductivity of the pore-network model without the trapped gas bubbles for mono-sized and distributed-size cases (second row in Fig. 4). For the mono-sized case, the normalized hydraulic conductivity starts decreasing with increasing gas bubbles noticeably at the bubble radius $R_B = 20$μm which is near at the average of tube radius. For the case of the initial gas bubble radius larger than $R_B = 35$μm, the value of the reduced hydraulic conductivity is almost constant depending on the N_B/N_P ratio. Especially for the case of the mono-sized gas bubble larger than $R_B = 35$μm and $N_B/N_P = 0.8$ case, the hydraulic conductivity of the pore-network model at the end of the simulation becomes zero even at $S_r = 0.07$. For the mono-sized $R_B = 50$μm and $N_B/N_P = 0.2$ case, the hydraulic conductivity is reduced to 49% of the initial hydraulic conductivity at the very low gas saturation Sg = 0.07. However, for the distributed-sized case, the hydraulic conductivity at the end of the simulation is reduced gradually as the R_B/R_P ratio increases, and the reduced values becomes constant for the R_B/R_P ratio higher than $R_B/R_P = 0.33$.

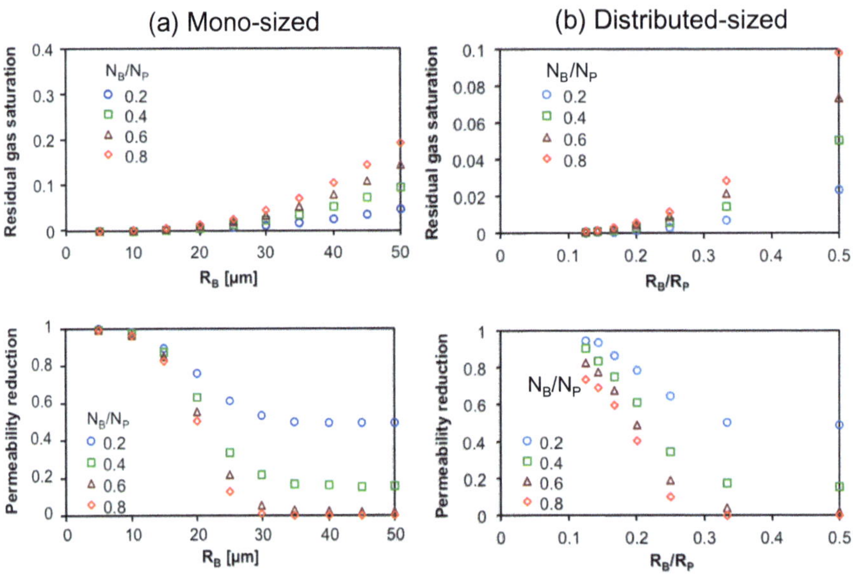

Fig. 4. Total migration time, trapped bubble fraction, residual saturation, and permeability reduction for (a) Mono-sized bubble migration and (b) Distributed-size bubble migration.

Relevance to in situ condition. In this study, all gas bubbles are generated instantaneously in the beginning of the simulation, and then the bubbles start migrating. However, the gas generation rate will be dependent on the in situ condition: The gas bubbles are generated very slowly via the natural microbial activity (Abrams 2005; Leifer and Patro 2002; Whalen 2005). The gas bubble generation by the denitrification process can be facilitated by injecting nutrient and controlling the environment such as the pH value of pore fluid (He and Chu 2014; Rebata-Landa and Santamarina 2012). Sometimes, gas bubbles can be generated very rapidly via the depressurization in the methane hydrate-bearing reservoir (Jang and Santamarina 2014).

The gas bubble generation rate and the initial gas bubble size upon nucleation can affect the gas bubble migration and the trapping. The gas bubble size upon nucleation and slow bubble generation rate (corresponding to the small R_B size and the low N_B/N_P case used in this study) will make it easier for gas bubbles to migrate upward without having coalescence and trapping. However, the large size of gas bubbles upon nucleation and rapid gas bubble generation rate (corresponding to the large R_B size and the high N_B/N_P case in this study) will facilitate the coalescence of gas bubbles and the possibility of trapping increases. Therefore, the in situ gas generation rate should be considered in order to apply the results of the pore-network model simulation to analyze the in situ gas bubble behavior.

The gas bubble movement (e.g., migration and trapping) through the porous media (which are in the sinusoidal shape consisting of wide pores and narrow pore throats) could be different than the gas bubble movement in the cylindrical tubes (Roosevelt and Corapcioglu 1998). A gas bubble whose size is equivalent to the pore size could migration and pass through the pore throat. Therefore, the pore-network model simulation needs to be carefully used to understand the gas bubble behavior (e.g., gas bubble stability in pore space) at the in situ condition for a long-term prediction (e.g., effect of gas bubble formation on hydraulic conductivity and liquefaction prevention).

4 Conclusions

The coalescence of gas bubbles during the gas migration induces the change in the bubble size distribution. For both mono-sized and distributed-size cases, the bubble size distribution becomes wider and the total number of gas bubbles decreases during the simulation.

The residual gas saturation at the end of the simulation increases as the bubble size R_B for the mono-sized case and the R_B/R_P ratio for the distributed-sized case increases. This trend is more pronounced as the N_B/N_P ratio increases from $N_B/N_P = 0.2$ to 0.8. The hydraulic conductivity decreases due to the gas bubble trapping in the pore-network model. The hydraulic conductivity at the end of the simulation decreases as the gas bubble size R_B or the ratio R_B/B_P increases. The reduction of hydraulic conductivity is significant for the N_B/N_P ratio higher than 0.6: The hydraulic conductivity becomes zero for $R_B > 35 \mu m$.

Finally, the gas bubble behavior and the associated property change obtained by the pore-network model simulation needs to be carefully used to predict the in situ gas bubble behavior due to the assumption used in this study.

Acknowledgements. This work was supported by the research fund of Hanyang University (HY-201700000002411).

References

Abrams, M.A.: Significance of hydrocarbon seepage relative to petroleum generation and entrapment. Mar. Pet. Geol. **22**, 457–477 (2005). https://doi.org/10.1016/j.marpetgeo.2004.08.003

Amos, R.T., Mayer, K.U.: Investigating ebullition in a sand column using dissolved gas analysis and reactive transport modeling. Environ. Sci. Technol. **40**, 5361–5367 (2006)

Amos, R.T., Mayer, K.U., Bekins, B.A., Delin, G.N., Williams, R.L.: Use of dissolved and vapor-phase gases to investigate methanogenic degradation of petroleum hydrocarbon contamination in the subsurface. Water Resour. Res. **41**(W02001), 02001–02015 (2005). https://doi.org/10.1029/2004WR003433

ASTM.: D7928-16 Standard test method for particle-size distribution (gradation) of fine-grained soils using the sedimentation (hydrometer) analysis, edited (2016)

Bora, R., Maini, B.B., Chakma, A.: Flow visualization studies of solution gas drive process in heavy oil reservoirs with a glass micromodel. SPE Reservoir. Eval. Eng. **3**(3), 224–229 (2000)

Dong, H., Blunt, M.: Pore-network extraction from micro-computerized-tomography images. Phys. Rev. E **80**(3), 0001 (2009). https://doi.org/10.1103/PhysRevE.80.036307

Enouy, R., Li, M., Ioannidis, M.A., Unger, A.J.A.: Gas exsolution and flow during supersaturated water injection in porous media: II Column experiments and continuum modeling. Adv. Water Res. **34**, 15–25 (2011). https://doi.org/10.1016/j.advwatres.2010.09.013

Eseller-Bayat, E., Yegian, M.K., Alshawabkeh, A., Gokyer, S.: Liquefaction Response of Partially Saturated Sands. I: Experimental Results. J. Geotechn. Geoenviron. Eng. **139**(6), 863–871 (2013). https://doi.org/10.1061/(asce)gt.1943-5606.0000815

Grozic, J.L., Robertson, P.K., Morgenstern, N.R.: The behavior of loose gassy sand. Can. Geotech. J. **36**, 482–492 (1999)

He, J., Chu, J.: Undrained Responses of Microbially Desaturated Sand under Monotonic Loading. J. Geotechn. Geoenviron. Eng. **140**(5), 04014003 (2014). https://doi.org/10.1061/(asce)gt.1943-5606.0001082

Jang, J., Santamarina, J.C.: Recoverable gas from hydrate-bearing sediments: Pore network model simulation and macroscale analyses. J. Geophys. Res. **116**(B8), 0001 (2011). https://doi.org/10.1029/2010jb007841

Jang, J., Santamarina, J.C.: Evolution of gas saturation and relative permeability during gas production from hydrate-bearing sediments: Gas invasion vs. gas nucleation. J. Geophys. Res. Solid Earth **119**(1), 116–126 (2014). https://doi.org/10.1002/2013jb010480

Jang, J., Santamarina, J.C.: Hydrate bearing clayey sediments: formation and gas production concepts. Mar. Pet. Geol. **77**, 235–246 (2016). https://doi.org/10.1016/j.marpetgeo.2016.06.013

Jang, J., Narsilio, G.A., Santamarina, J.C.: Hydraulic conductivity in spatially varying media-a pore-scale investigation. Geophys. J. Int. **184**(3), 1167–1179 (2011). https://doi.org/10.1111/j.1365-246X.2010.04893.x

Krol, M.M., Mumford, K.G., Johnson, R.L., Sleep, B.E.: Modeling discrete gas bubble formation and mobilization during subsurface heating of contaminated zones. Adv. Water Resour. **34**, 537–549 (2011). https://doi.org/10.1016/j.advwatres.2011.01.010

Leifer, I., Patro, R.K.: The bubble mechanism for methane transport from the shallow sea bed to the surface: A review and sensitivity study. Cont. Shelf Res. **22**, 2409–2428 (2002)

Mahabadi, N., Jang, J.: Relative water and gas permeability for gas production from hydrate-bearing sediments. Geochem. Geophys. Geosyst. **15**, 2346–2353 (2014). https://doi.org/10.1002/2014GC005331

Mahabadi, N., Zheng, X., Jang, J.: The effect of hydrate saturation on water retention curves in hydrate-bearing sediments. Geophys. Res. Lett. **43**(9), 4279–4287 (2016a). https://doi.org/10.1002/2016gl068656

Mahabadi, N., Dai, S., Seol, Y., Yun, T.S., Jang, J.: The water retention curve and relative permeability for gas production from hydrate-bearing sediments: pore-network model simulation. Geochem. Geophys. Geosyst. **17**, 3099–3110 (2016b). https://doi.org/10.1002/2016GC006372

McCray, J.E., Falta, R.W.: Numerical simulation of air sparging for remdiation of NAPL contamination. Ground Water **35**(1), 99–110 (1997)

Moskvitch, K.: Mysterious Siberian crater attributed to methane. Nature (2014). https://doi.org/10.1038/nature.2014.15649

Plampin, M., Illangasekare, T., Sakaki, T., Pawar, R.: Experimental study of gas evolution in heterogeneous shallow subsurface formations during leakage of stored CO2. Int. J. Greenhouse Gas Control **22**, 47–62 (2014). https://doi.org/10.1016/j.ijggc.2013.12.020

Ramirez, J.A., Baird, A.J., Coulthard, T.J., Waddington, J.M.: Testing a simple model of gas bubble dynamics in porous media. Water Resour. Res. **51**(2), 1036–1049 (2015). https://doi.org/10.1002/2014wr015898

Rebata-Landa, V., Santamarina, J.C.: Mechanical Effects of Biogenic Nitrogen Gas Bubbles in Soils. J. Geotech. Geoenviron. Eng. **138**(2), 128–137 (2012). https://doi.org/10.1061/(asce)gt.1943-5606.0000571

Ronen, D., Berkowitz, B., Magaritz, M.: The development and influence of gas bubbles in phreatic aquifers under natural flow conditions. Transp. Porous Media **4**, 295–306 (1989)

Roosevelt, S.E., Corapcioglu, M.Y.: Air bubble migration in a granular porous medium: Experimental studies. Water Resour. Res. **34**(5), 1131–1142 (1998). https://doi.org/10.1029/98wr00371

Ryan, M.C., MacQuarrie, K.T.B., Harman, J., McLellan, J.: Field and modeling evidence for a "stagnant flow" zone in the upper meter of sandy phreatic aquifers. J. Hydrol. **233**, 223–240 (2000)

Sills, G.C., Wheeler, S.J., Thomas, S.D., Gardner, T.N.: Behaviour of offshore soils containing gas bubbles. Geotechnique **41**(2), 227–241 (1991)

Stewart, C.R., Hunt Jr., E.B., Schneider, F.N., Geffen, T.M., Berry Jr., V.J.: The role of bubble formation in oil recovery by solution gas drives in limestones. J. Petrol. Technol. **5**(12), 21–28 (1954)

van Breukelen, B.M., Roling, W.F.M., Groen, J., Griffioen, J., van Verseveld, H.W.: Biogeochemistry and isotope geochemistry of a landfill leachate plume. J. Contam. Hydrol. **65**, 245–268 (2003). https://doi.org/10.1016/S0169-7722(03)00003-2

Walter, K.M., Zimov, S.A., Chanton, J.P., Verbyla, D., Chapin 3rd, F.S.: Methane bubbling from Siberian thaw lakes as a positive feedback to climate warming. Nature **443**(7107), 71–75 (2006). https://doi.org/10.1038/nature05040

Whalen, S.C.: Biogeochemistry of methane exchange between natural wetlands and the atmosphere. Environ. Eng. Sci. **22**(1), 73–94 (2005)

Zheng, X., Jang, J.: Hydraulic Properties of Porous Media Saturated with Nanoparticle-Stabilized Air-Water Foam. Sustainability **8**(12), 1317 (2016). https://doi.org/10.3390/su8121317

Zheng, X., Mahabadi, N., Yun, T.S., Jang, J.: Effect of capillary and viscous force on CO_2 saturation and invasion pattern in the microfluidic chip. J. Geophys. Res. **122**, 1634–1647 (2017). https://doi.org/10.1002/2016JB013908

Zuo, L., Krevor, S., Falta, R.W., Benson, S.M.: An Experimental Study of CO_2 Exsolution and Relative Permeability Measurements During CO_2 Saturated Water Depressurization. Transp. Porous Media **91**(2), 459–478 (2012). https://doi.org/10.1007/s11242-011-9854-2

Zuo, L., Zhang, C., Falta, R.W., Benson, S.M.: Micromodel investigations of CO_2 exsolution from carbonated water in sedimentary rocks. Adv. Water Resour. **53**, 188–197 (2013). https://doi.org/10.1016/j.advwatres.2012.11.004

A Coupled Chemo-Mechanical Analysis of the Dissolution-Dominated Sinkholes

Suraj Khadka and Liang-Bo Hu(✉)

Department of Civil Engineering, University of Toledo, Toledo, OH 43606, USA
Liangbo.Hu@utoledo.edu

Abstract. Sinkholes pose a major threat to public safety and infrastructure. They can develop via a cluster of inter-related processes, including bedrock dissolution, rock collapse, soil washing and soil collapse. The dominant mechanism behind sinkholes formed in rocks is the dissolution of soluble rocks. Dissolution process may be enhanced by potentially aggressive groundwater acidity and the presence of caves or fissures. This paper presents a coupled chemo-mechanical approach to understanding the interaction of chemical reaction and mechanical deformation processes involved in sinkhole development. Dissolution kinetics and enhanced deformation processes are investigated. Specific solution rate of the constituent mineral and the surface area available for the reaction are related via a chemo-mechanical coupling with the consideration of the damage-enhanced dissolution mechanism. Another important coupling investigated is the potential weakening of rock materials due to dissolution. Kinetic rates of different minerals are surveyed and used to examine the dissolution enhanced deformation. Boundary value problems are formulated around the cavity to simulate the progression of mineral dissolution and plastic deformation.

1 Introduction

Sinkholes typically develop in karstic rock masses or soil sediments overlying the rock masses (Wilson and Beck 1992; Tharp 1999; Waltham et al. 2005; Parise and Lollino 2011; Gutierrez et al. 2014). These soluble rocks primarily refer to carbonate rocks (limestone, dolomite) and gypsum. Limestone is one of the world's most widespread sedimentary rocks, and karst is developed to some degree in almost every country of the world. Sinkholes can develop by a cluster of inter-related processes (e.g., Gutierrez et al. 2014), including bedrock dissolution, rock collapse, soil down-washing and soil collapse.

Sinkholes can be classified in two major categories, sinkholes formed in the karstic rock (limestone, gypsum or salt), and those formed in soils overlying the karstic rocks (Waltham et al. 2005). The dominant process behind sinkholes formed in rocks is the dissolution of soluble rocks. Dissolution process may be enhanced by potentially aggressive groundwater acidity and result in various dissolution features such as caves or fissures (Wang et al. 2017); these caves or fissures, in turn, may continue to expand or grow as a result of dissolution. Collapse and caprock sinkholes are defined by fracturing, breakdown, and collapse of bedrock slabs and arches as they gradually lose

S. Wang et al. (eds.), *New Solutions for Challenges in Applications of New Materials and Geotechnical Issues*, Sustainable Civil Infrastructures,
https://doi.org/10.1007/978-3-319-95744-9_4

the support around dissolution cavities. Sinkholes formed in soils are a widespread geo-hazard. They are generally caused by the erosion, transport, and failure of the soils that overlie cavernous rock. Because of its low strength compared to rock, which, if left over a cave can still be strong enough to stand for a long period of time, a soil arch over a void is inherently unstable and its collapse can occur rapidly. Their underlying mechanisms are extremely intricate and traditionally have received more attention from the geotechnical communities (e.g., Tharp 1999; Goodings and Abdulla 2002; Augarde et al. 2003), including some of our earlier work (Rawal et al. 2016, 2017).

The present study is focused on the dissolution-dominated sinkholes which mainly occur in karst rocks. Although they are a less widespread hazard than sinkholes in soils and pose a lower catastrophic threat, a better understanding of the involved mechanisms is still of vital importance. Even though the geochemical reactions in the karstic carbonate rocks are generally very slow and their hazardous consequences like sinkholes occurrence, caves and fissures formation may usually take extensive periods of time, typically in geological scale, increased human activities in the society have caused many dramatic effects, many of which lead to accelerated or intensified chemical interactions; for example aggressive groundwater with high acidity can substantially enhance the dissolution of limestone or dolomite. In addition, dissolution of gypsum and salt takes place fairly quickly, the latter has led to increased sinkholes around the shore of Dead Sea (Shalev et al. 2006).

The dissolution dominated sinkholes are inherently related to the development of karst environments typically characterized by distinctive landforms arising from dissolution and subsurface drainage. Intriguing dynamics between the mineral dissolution and karst evolution, as well as various hazards associated with karst environments have been intensively studied by research communities (e.g., White 1988; White 2002; Ford and Williams 2007; Parise and Gunn 2007; Zhou and Beck 2008; Gutierrez 2010; Parise 2011; De Waele et al. 2011; Gutierrez et al. 2014). Of particular interest is the geotechnical modeling of the hydrological and mechanical processes involved in sinkholes. Kaufmann and coworkers have developed a numerical package to simulate the evolution of karst aquifer and collapse sinkhole (Kaufmann et al. 2010; Hiller et al. 2011; Kaufmann and Romanov 2016). Ghabezloo and Pouya (2006) developed a numerical chemical model for the assessment of the progression of weathering process within the karst caves, which highlighted the need of investigation of the effect of carbonate dissolution and related mechanical degradation in time-dependent progression of rock-mass failures around the underground cave. Shalev and Lyakhovsky (2012) used a viscoelastic damage rheology model to simulate the rock behavior around a cavity in sinkhole formation. Parise and Lollino (2011) and Lollino et al. (2013) developed a numerical approach to model the stability of failure in underground caves where the degradation of the rock mass potentially due to environmental weathering was considered.

In the present numerical study, the coupled processes of deformation and dissolution in karstic rocks are investigated in a chemo-plasticity framework. Kinetic rates of several typical karst rock minerals are surveyed. Subsequently, the enhanced deformation induced by the chemical dissolution is investigated at a closed-system level. Finally, typical sinkhole scenarios formulated in a Boundary Value Problem (BVP) setting are numerically examined.

2 Mechanisms and Modelling Framework

In this section, we present a general framework for the coupled processes of deformation and dissolution in karst rocks, followed by numerical examples of the typical kinetic rate and dissolution enhanced deformation. Theoretical developments will adopt the classical formulations of dissolution kinetics for limestone (calcite and dolomite), gypsum and salt. The coupling effect is addressed by the specific surface area affected by mechanical deformation.

The dissolution enhanced damage can be addressed in constitutive formulations for karst rock minerals, considering the chemical changes and strength evolution. To better illustrate the possible approach, a chemo-plasticity model is presented here, assuming rigid plasticity. The yielding function can be generally defined as

$$f = f\left(\sigma_j, p_c\right) = 0 \tag{1}$$

$p_c = p_c\left(\varepsilon_q^{pl}, \xi\right)$ describes an isotropic size characteristic of yield locus (i.e., strength). It depends on a set of hardening and/or softening parameters that are either mechanical or chemical in nature. ε_q^{pl} is the deviatoric strain hardening parameter, defined as

$$\varepsilon_q^{pl} = \left(\frac{2}{3}\dot{e}_{ij}^{pl}\dot{e}_{ij}^{pl}\right)^{1/2}, \quad \dot{e}_{ij}^{pl} = \varepsilon_{ij}^{pl} - \frac{1}{3}\dot{\varepsilon}_{kk}^{pl}\delta_{ij} \tag{2}$$

Here the chemical parameter, ξ is chosen to be the mass removal of the material, its evolution is described by the dissolution kinetic rate which will be discussed in the next section, as we can define

$$\xi = \Delta m / m_0 \tag{3}$$

It is the ratio of change of mass (or mol) to the original mass (or mol) of the mineral, thus it is confined to the range [0, 1] and can be directly used as a softening parameter.

This framework reflects two different and independent ways in which the material may become harder or softer. One is a classical deviatoric strain hardening and the other reflects the removal of mass in the weakening of the material. The following derivation can be easily obtained based on the associated flow rule and Prager's consistency condition,

$$\dot{\varepsilon}_{ij}^{pl} = -\frac{1}{H}\frac{\partial f}{\partial \sigma_{ij}}\left(\frac{\partial f}{\partial \sigma_{kl}}\dot{\sigma}_{kl} + \frac{\partial f}{\partial \xi}\dot{\xi}\right), \text{ where } H = \frac{\partial f}{\partial \varepsilon_q^{pl}}\left[\frac{2}{3}\frac{\partial f}{\partial s_{ij}}\frac{\partial f}{\partial s_{ij}}\right]^{\frac{1}{2}}, s_{ij} = \sigma_{ij} - \frac{1}{3}\sigma_{kk}\delta_{ij} \tag{4}$$

It describes the dependence of plastic strain on stress (mechanical loading) and dissolution (chemical effect). The couplings summarized above can be used to model the deformation and dissolution of karstic rocks, complementing the stress equilibrium and kinematic relationships for formulated boundary value problems (BVPs).

In the following, we first attempt to survey the different kinetic rates of common karst rock minerals and subsequently demonstrate the dissolution enhanced deformation via the chemo-mechanical coupling discussed above.

2.1 Typical Dissolution Kinetic Rates in Karst Rocks

The rate of dissolution of rocks is believed to generally depend on the solubility and specific solution rate constant of the constituent mineral, and the surface area available for the reaction to occur, in an over-simplification of the potentially very complex reaction processes and kinetics. In the present study, we focus on four typical karst rock minerals: calcite, dolomite, gypsum, and salt (halite). The first two are the most common minerals forming rocks in karst terrain and their rates are much slower than the latter two; gypsum has a moderately fast dissolution and salt dissolves very rapidly under normal environmental conditions. For comparison, the following rate equation is adopted for all minerals,

$$\dot{m} = kA\left(1 - \frac{C}{C^{\mathrm{sat}}}\right) \tag{5}$$

The dissolution rate, $\dot{m}$ (mol/s), as also related to Eqs. (3) and (4), is dependent on the specific surface area, A ($\mathrm{m}^2/\mathrm{m}^3$). k is the rate constant ($\mathrm{mol/m^2/s}$). C and C^{sat}($\mathrm{mol/m^3}$) are the concentration and the equilibrium (saturation) concentration of the mineral, respectively. Obviously C^{sat} can be related to the so-called solubility, S ($\mathrm{kg/m^3}$), via $S = C^{\mathrm{sat}} v_m$, v_m is the molar mass of the mineral.

We can simulate the dissolution around a cavity filled with water. Considering a closed system regarding the mineral concentration in the cavity (R = 1 m) where all dissolved mineral remains in the cavity, it can be established that $\dot{C} = \dot{m}/V = kA/V(1 - C/C^{\mathrm{sat}})$, V is the volume of the cavity; integration of the equation directly leads to:

$$C = C^{\mathrm{sat}}\left(1 - \exp\left[-\frac{kA}{VC^{\mathrm{sat}}}t\right]\right) \tag{6}$$

Subsequently, the depletion of the mineral dissolved around the cavity can be computed as

$$d = \frac{v_m C^{\mathrm{sat}} V}{\rho A}\left(1 - \exp\left[-\frac{kA}{VC^{\mathrm{sat}}}t\right]\right) \tag{7}$$

ρ is the density of the mineral. The solution concentration and the depletion are shown in Fig. 1. The major constants used in the calculations are summarized in Table 1, where the rate constants and solubility were based on the data reported in Ford and Williams (2007). Figure 1a shows that salt and gypsum dissolve very rapidly, taking a few minutes and hours, respectively, while it would take calcite and dolomite hundreds of hours to reach equilibrium. Due to the much higher solubility, salt and gypsum deplete substantially as a result of surface dissolution around the cavity as shown in Fig. 1b. The depletion of calcite and dolomite is quite modest, of course, under the assumed specific scenario of mineral concentration growing in the solution water in the cavity to eventually reach saturation. In actual geological masses the surrounding solution varies during the hydrological processes, the dissolution process would progress continuously once the solution is under the saturation concentration; this is explored in the following section.

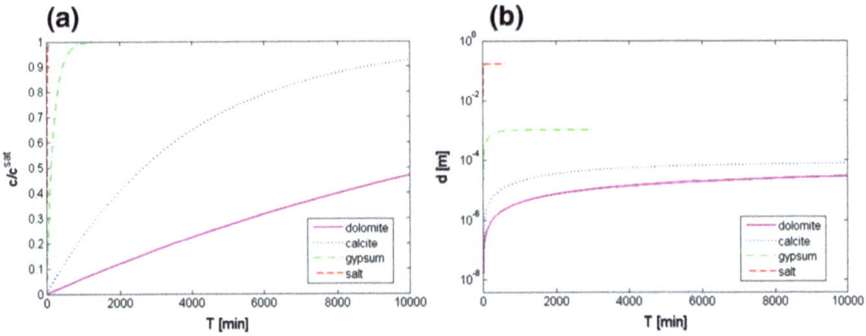

Fig. 1. (a) Concentration evolution of mineral solution in the cavity. (b) Depletion of mineral around the cavity.

Table 1. Constants used in the simulation.

Mineral	Density (g/cm^3)	Molar mass (g/mol)	Rate constant ($mmol/cm^2/s$)	Solubility (mg/L)
Dolomite	2.85	184	1×10^{-7}	175
Calcite	2.71	100	1×10^{-6}	230
Gypsum	2.32	172	1.2×10^{-4}	2,400
Salt	2.16	58	6.7×10^{-1}	360,000

2.2 Dissolution Enhanced Deformation

A critical parameter of the geochemical kinetic law is the specific surface area, A. As mechanical damage induces micro-cracking leading to a new surface area to be opened, this parameter is then lined with mechanical strain. For example, the cumulative frequency of acoustic emission events is known to be proportional to the inelastic dilatant volumetric strain (Brace et al. 1966; Scholz 1968). Hence, the new interface surface area generated by the micro-cracking per unit volume of the medium has been proposed to be dependent on plastic dilatancy using a single model of a two-dimensional hexagonal crystal assembly (Hu and Hueckel 2007). It is proposed that the specific surface area is dependent on the plastic volumetric strain, ϵ_v^{pl} via, $A = A\left(\epsilon_v^{pl}\right) = \gamma\epsilon_v^{pl} + A_0$. A_0 describes the original available specific surface for dissolution.

The above formulation is used in the presented numerical example of dissolution enhanced deformation. We would attempt to first assess a closed-system, element level behavior when the mineral is subject to dissolution and under a state of constant stress, which is typical of situations where a constant load (surcharge) is applied on a geological field or mass, and the stress at any material point varies very little despite that the mineral is subject to dissolution enhancing the deformation or damage. This scenario leads a simplification of Eq. (4), the first term of the left-handed side varnishes owing to $\dot{\sigma}_{ij} = 0$, a creep-like strain develops.

The plasticity model employed is based on the formulation proposed by Dragon and Mróz (1979):

$$f\left(\sigma_{ij}, \kappa, \xi\right) = s_{ij}s_{ij} - 2p_h\left(I_1^0 - \sigma_{kk}\right) \tag{8}$$

where s_{ij} is the deviatoric stress, while the hardening parameter, p_{h}, defining the size of the yield locus, is a resultant of a mechanical and chemical hardening.

$$p_h(\kappa, \xi) = p_0\left(1 + \alpha_0\kappa - \beta_0 \mathrm{In}\frac{\beta_1 - \xi_0}{\beta_1 - \zeta}\right) \tag{9}$$

α_0 and β_0 are material constants. β_1 is a constant defining the chemical softening curve. α and β are two independent hardening functions. I_1^0 is a constant isotropic stress value at which the yield limit intersects the isotropic stress axis. Parameter $p_h = p_0$ at no hardening is related to the pre-peak yielding stress value. The details of the mathematical formulation can be found in Hu and Hueckel (2007). Based on Eq. (4) and noting that $\dot{\sigma}_{ij} = 0$, the volumetric strain can be readily established as

$$\dot{\epsilon}_v = \frac{\partial p_h/\partial \xi}{\partial p_h/\partial \kappa}\frac{p_h\delta_{kk}}{\left(\frac{2}{3}s_{kl}s_{kl}\right)^{1/2}}\dot{\xi} = \frac{\beta_0}{(\beta_1-\xi)\alpha_0}\frac{3p_h}{\left(\frac{2}{3}s_{kl}s_{kl}\right)^{1/2}}\dot{\xi} \tag{10}$$

The above equation is complemented by the rate of the mineral mass loss ratio, $\dot{\xi} = \dot{m}/m_0$, based on the kinetic rate equation discussed earlier,,

$$\dot{\xi} = \frac{v_m k(\gamma\epsilon_v^{pl} + A_0)}{\rho} \tag{11}$$

Equations (10) and (11) form a pair of ODEs that can be solved under any given initial conditions. It is also assumed that the concentration in the cavity is immediately carried away and the solution remains under-saturation all the time, thus the dissolution continues until the mineral is completely depleted.

In the present simulation, we also consider that the dissolution of limestone or dolomite in pure water is extremely low but can be accelerated considerably in acidic conditions (Sjoberg 1976; Sjoberg and Rickard 1984) under which the dissolution rate of calcite could rise over several orders of magnitude. Hence a modest range of acidity and the associated kinetic rate is explored in the simulation as shown in Fig. 2. A proportional relationship between the rate constant and [H^+] concentration (note that pH = log [H^+]) is used (Sjoberg 1976; Ciantia and Hueckel 2013). The strain developed in the calcite mineral is solely induced by the mineral dissolution under constant stresses as described by Eq. (10). As shown in Fig. 2, under neutral or low acidity the growth develops rather slow and is not significant even after a very long period of time. Strong acidity drastically accelerates the dissolution and simultaneously the deformation can reach a very high level after a few days. The evolution exhibits a strong linear relationship till toward the end of dissolution.

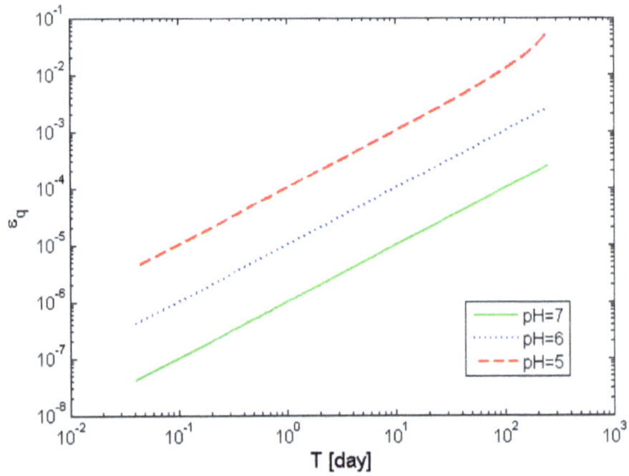

Fig. 2. Development of deviatoric strain induced by the calcite mineral dissolution at various levels of acidity.

3 Numerical Simulations of Cavity Deformation

Besides element level modeling, numerical simulations can also be performed in the boundary value problems (BVPs) in which both the evolution and distribution of deformation can be assessed. Figure 3 shows the domain around a cavity (R = 1 m) subjected to chemical dissolution, under a constant load from the top surface, established in the computational software FLAC (Itasca 2011). The chemical softening mechanism is introduced to the constitutive law for the simulated rock, a conventional Mohr-Coulomb failure function is modified with the shear strength parameters (e.g., cohesion) decreasing as a result of chemical dissolution.

Figure 4 shows the progressive development of the settlement at the top of the cavity, under different levels of acidity attacking the soluble calcite around the cavity. The dramatic rise in the enhanced deformation after certain periods of time is accompanied with the propagation in the plasticity zone around the cavity. It is evident that faster dissolution weakens the mineral more rapidly and induces more significant deformation. For simplicity, in the presented study a constant dissolution rate is used by ignoring the strain-dependent term in eq. (11), more intricate formulations can be also explored.

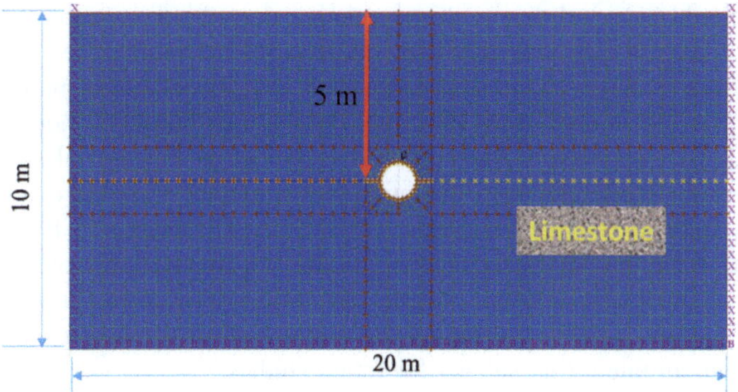

Fig. 3. Geometry of the simulated limestone formation around a cavity of 2-m in diameter.

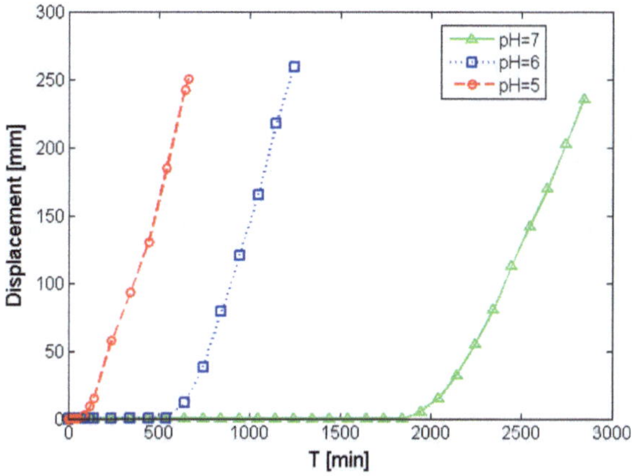

Fig. 4. Progression of the displacement at the top of the cavity under different levels of acidity.

4 Conclusions

The present study is primarily focused on the mechanisms of dissolution-dominated sinkholes. A coupled chemo-mechanical model is applied. Kinetic rates of different minerals are examined numerically in the context of a cavity dissolution, where the examined scenarios consider the dynamic equilibrium between the dissolution and precipitation processes around the cavity. Actual scenarios in the field typically involve the transport of dissolving minerals and water flow; consideration of reactive transport of hydrogeological processes may be necessary when evaluating the dissolution processes at a large field scale. The kinetic rates of different minerals are subsequently used to demonstrate the dissolution enhanced damage for a closed system which evolves with time under a constant level of stresses. While their effects are less rapid and less strong than fast dissolution, slow dissolutions in carbonate rocks can still induce significant deformation and the effects can accumulate over prolonged periods of time. It should be noted that in this simulation only the dissolution is considered, reflective of a fastest case scenario when no precipitation is produced to counter the mineral dissolution. BVPs can be also formulated and solved with computational tools to simulate possible sinkhole scenarios after the coupled plasticity model is properly implemented. The influence of kinetic rate is examined at different pH levels. Strong acidity drastically accelerates the dissolution and simultaneously the deformation can reach a very high level after a short period of time. The presented BVP simulation considers the softening of material strength due to chemical dissolution which is assumed to progress at a constant kinetic rate. Many intricacies about the complex evolution of fissure or fracture opening due to mechanical damage for enhanced dissolution remain an interesting subject for future studies.

References

Augarde, C.E., Lyamin, A.V., Sloan, S.W.: Prediction of undrained sinkhole collapse. J. Geotechn. Geoenviron. Eng. **129**(3), 197–205 (2003). https://doi.org/10.1061/(ASCE)1090-0241

Brace, W.F., Paulding, B.W., Scholz, C.: Dilatancy in the fracture of crystalline rocks. J. Geophys. Res. **71**, 3939–3953 (1966). https://doi.org/10.1029/JZ071i016p03939

Ciantia, M.O., Hueckel, T.: Weathering of submerged stressed calcarenites: chemo-mechanical coupling mechanisms. Geotechnique **63**(9), 768–785 (2013). https://doi.org/10.1680/geot.SIP13.P.024

De Waele, J., Gutiérrez, F., Parise, M., Plan, L.: Geomorphology and natural hazards in karst areas: a review. Geomorphology. **134**, 1–8 (2011). https://doi.org/10.1016/j.geomorph.2011.08.001

Dragon, A., Mróz, Z.: A model for plastic creep of rock-like materials accounting for the kinetics of fracture. Int. J. Rock Mech. Min. Sci. **16**, 253–259 (1979). https://doi.org/10.1016/0148-9062(79)91200-2

Itasca: FLAC2D User's Guide, 5th edn. Itasca Consulting Group Inc., Minneapolis, USA (2011)

Ford, D., Williams, P.: Karst Hydrogeology and Geomorphology. Wiley, West Sussex, UK (2007)

Ghabezloo, S., Pouya, A.: Numerical Modelling of the effect of weathering on the progressive failure of underground limestone mines. In: Eurock 2006, Multiphysics Coupling and Long Term Behaviour in Rock Mechanics: Proceedings of the International Symposium of the International Society for Rock Mechanics, Liège, Belgium (2006). https://doi.org/10.1201/9781439833469.ch32

Goodings, D.J., Abdulla, W.A.: Stability charts for predicting sinkholes in weakly cemented sand over karst limestone. Eng. Geol. **65**(2), 179–184 (2002). https://doi.org/10.1016/S0013-7952(01)00127-2

Gutiérrez, F.: Hazards associated with karst. In: Alcántara I, Goudie A (Eds), Geomorphological Hazards and Disaster Prevention. Cambridge University Press, Cambridge, 161–175 (2010). https://doi.org/10.1017/cbo9780511807527.013

Gutierrez, F., Parise, M., De Waele, J., Jourde, H.: A review on natural and human-induced geohazards and impacts in karst. Earth-Sci. Rev. **138**, 61–88 (2014). https://doi.org/10.1016/j.earscirev.2014.08.002

Hiller, T., Kaufmann, G., Romanov, D.: Karstification beneath dam sites: from conceptual models to realistic scenarios. J. Hydrol. **398**, 202–211 (2011). https://doi.org/10.1016/j.jhydrol.2010.12.014

Hu, L.B., Hueckel, T.: Creep of saturated materials as a chemically enhanced rate dependent damage process. Int. J. Numer. Anal. Meth. Geomech. **31**(14), 1537–1565 (2007). https://doi.org/10.1002/nag.600

Kaufmann, G., Romanov, D., Hiller, T.: Modelling three-dimensional karst aquifer evolution using different matrix-flow components. J. Hydrol. **388**, 241–250 (2010). https://doi.org/10.1016/j.jhydrol.2010.05.001

Kaufmann, G., Romanov, D.: Structure and evolution of collapse sinkholes: Combined interpretation from physico-chemical modelling and geophysical field work. J. Hydrol. **540**, 688–698 (2016). https://doi.org/10.1016/j.jhydrol.2016.06.050

Lollino, P., Martimucci, V., Parise, M.: Geological survey and numerical modeling of the potential failure mechanisms of underground caves. Geosystem Engineering **16**(1), 100–112 (2013). https://doi.org/10.1080/12269328.2013.780721

Parise, M.: Surface and subsurface karst geomorphology in the Murge (Apulia, southern Italy). Acta Carsologica **40**(1), 79–93 (2011). https://doi.org/10.3986/ac.v40i1.30

Parise, M., Gunn, J.: Natural and anthropogenic hazards in karst areas: recognition, analysis and mitigation. Geolog. Soc. London, Spec. Publ. **279**, 1–3 (2007). https://doi.org/10.1144/SP279.10305-8719/07

Parise, M., Lollino, P.: A preliminary analysis of failure mechanisms in karst and man-made underground caves in southern Italy. Geomorphology **134**(1), 132–143 (2011). https://doi.org/10.1016/j.geomorph.2011.06.008

Rawal, K., Wang, Z.M., Hu, L.B.: Exploring the geomechanics of sinkholes: a numerical study of sinkhole subsidence and collapse. In: ASCE Geotechnical Special Publication 257: Proceedings of the 4th Geo-China International Conference, 1–8 (2016). https://doi.org/10.1061/9780784480007.001

Rawal, K., Hu, L.B., Wang Z.M.: Numerical investigation of geomechanics of sinkholes formation and subsidence. In: ASCE Geotechnical Special Publication 257: Proceedings of Geotechnical Frontiers 2017, 480–487 (2017). https://doi.org/10.1061/9780784480441.050

Scholz, C.H.: Microfracturing and the inelastic deformation of rock in compression. J. Geophys. Res. **73**(4), 1417–1432 (1968). https://doi.org/10.1029/JB073i004p01417

Shalev, E., Lyakhovsky, V., Yechieli, Y.: Salt dissolution and sinkhole formation along the Dead Sea shore. J. Geophysical Res. Solid Earth **111**, B03102 (2006). https://doi.org/10.1029/2005JB004038

Shalev, E., Lyakhovsky, V.: Viscoelastic damage modeling of sinkhole formation, In Journal of Structural Geology, vol. 42. ISSN **163–170**, 0191–8141 (2012). https://doi.org/10.1016/j.jsg.2012.05.010

Sjoberg, E.L.: A fundamental equation for calcite dissolution kinetics. Geochim. Cosmochim. Acta **40**, 441–447 (1976). https://doi.org/10.1016/0016-7037(76)90009-0

Sjoberg, E.L., Rickard, D.T.: Temperature dependence of calcite dissolution kinetics between 1 and 62 °C at pH 2.7 to 8.4 in aqueous solutions. Geochim. Cosmochim. Acta **48**(3), 483–485 (1984). https://doi.org/10.1016/0016-7037(84)90276-X

Tharp, T.M.: Mechanics of upward propagation of cover-collapse sinkholes. Eng. Geol. **52**(1), 23–33 (1999). https://doi.org/10.1016/S0013-7952(98)00051-9

Waltham, T., Bell, F.G., Culshaw, M.: Sinkholes and Subsidence: Karst and Cavernous Rocks in Engineering and Construction. Springer-Verlag, Berlin (2005)

Wang, Z.M., Yang, G.L., Yang, R.D., Rawal, K., Hu, L.B.: Evaluating the factors influencing limestone-dissolution characteristics in the karst regions of Guizhou. China. J. Testing Evaluat. **45**(1), 220–229 (2017). https://doi.org/10.1520/JTE20160131

White, W.B.: Geomorphology and Hydrology of Karst Terrains. Oxford University Press, New York (1988)
White, W.B.: Karst hydrology: recent developments and open questions. Eng. Geol. **65**(2), 85–105 (2002). https://doi.org/10.1016/S0013-7952(01)00116-8
Wilson, W.L., Beck, B.F.: Hydrogeologic Factors Affecting New Sinkhole Development in the Orlando Area, Florida. Ground Water **30**, 918–930 (1992). https://doi.org/10.1111/j.1745-6584.1992.tb01575.x
Zhou, W., Beck, B.F.: Management and mitigation of sinkholes on karst lands: an overview of practical applications. Environ. Geol. **55**(4), 837–851 (2008). https://doi.org/10.1007/s00254-007-1035-9

Simplified Methodology for Stiffness Estimation of Double D Shaped Caisson Foundations

Pradeep Kumar Dammala[1,2(✉)], Saleh Jalbi[2], Subhamoy Bhattacharya[1,2], and Murali Krishna Adapa[1,2]

[1] University of Surrey, Guildford GU2 7XH, UK
pradeepkumardammala@gmail.com,
s.bhattacharya@surrey.ac.uk,
[2] Indian Institute of Technology Guwahati, Guwahati 781 039, India
salehjalbi@gmail.com, amurali@iitg.ernet.in

Abstract. Foundation stiffness plays a crucial role in the stability analysis of structures to incorporate the Soil Structure Interaction (SSI) effects. Current design approaches of estimating foundation stiffness include advanced dynamic finite element, distributed spring approach, and lumped spring approach. The aim of this paper is to overview the different methods for computing the foundation stiffness and to check their applicability. This has been done by considering an example: Saraighat Bridge supported on double D shaped caisson foundation. Advanced three dimensional finite element analysis is performed to extract the stiffness (Lateral, rotational, and coupling) of double D foundations and the results are compared to the representative circular foundations. It has been concluded that the stiffness of foundation can be significantly affected by its geometry. Furthermore, the stiffness functions are utilized in computing the fundamental frequency of the bridge and also compared with the frequency obtained from different approaches. The frequency estimated using the present study matches satisfactorily well with the monitored data testifying the validation of the work.

1 Introduction

Design engineers require the foundation stiffness for the stability analysis of structures to incorporate the Soil-Structure-Interaction (SSI) effects. The current available approaches in estimating the foundation stiffness include advanced finite/boundary element and Beams on Nonlinear Winkler Foundation (BNWF). Figure 1 schematically illustrates the approaches available for modelling the SSI in case of deep foundations. The BNWF approach (Fig. 1b) lacks the continuity of the soil while the earlier requires highly skilled expertise and is uneconomical. The lumped spring approach proposed by Poulus (1971)—refer to Fig. 1c, represents a simple method to idealize the foundation and surrounding soil using four different types of springs. This approach has been extensively used in dynamic analysis of structures such as offshore wind turbines (Arany et al. 2016; Shadlou and Bhattacharya 2016) and buildings (Gazetas 1991) in predicting the fundamental frequency. It requires the minimum input parameters to

S. Wang et al. (eds.), *New Solutions for Challenges in Applications of New Materials and Geotechnical Issues*, Sustainable Civil Infrastructures,
https://doi.org/10.1007/978-3-319-95744-9_5

estimate the stiffness values for the analysis helping the designer to arrive at a preliminary or approximate geometry of the foundation in a short period of time. In the lumped spring approach, non-dimensional solutions for the stiffness estimation of surface and deep foundations in various directions (vertical-K_V, lateral-K_L, rocking K_R and coupled lateral and rocking-K_{LR}) were proposed by many researchers (Poulus 1971; Banerjee and Davies 1978; Gazetas 1991; Carter and Kulhawy 1992; Higgins and Basu 2011; Shadlou and Bhattacharya 2016; Jalbi et al. 2017).

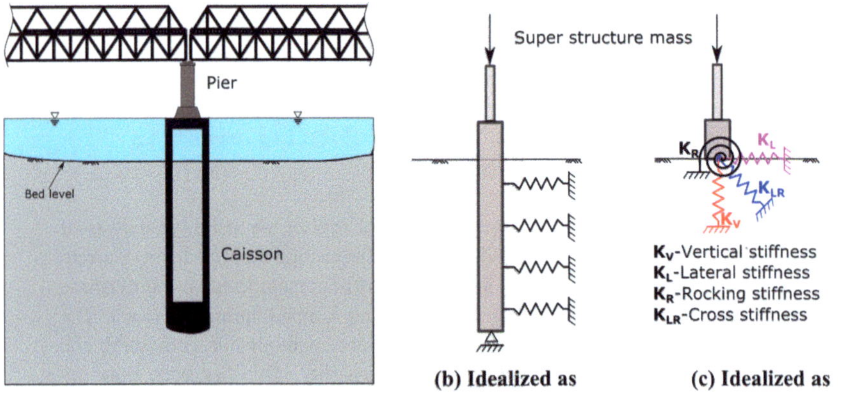

(a) Real field conditions **(b) Idealized as distributed soil springs** **(c) Idealized as lumped soil springs**

Fig. 1. Idealization of the foundation-soil interaction

Most of the studies are based on either a cylindrical or a rectangular/square shafts embedded in homogeneous/non homogeneous soil column. However, foundations with untraditional geometry may often be required to suit the purpose, such as bridge piers supporting multilane traffic decks in highly scouring rivers. In those cases, caissons of double D or hybrid combination of circular and rectangular sections are preferred, see for example Saraighat Bridge foundation (Dammala et al. 2017a). It is of interest to point out whether the established non-dimensional forms are sufficient enough to analyze an untraditionally shaped foundation or in a nutshell, do the available stiffness formulations take care of geometrical effects of the foundation?

1.1 Objectives

This article highlights the significance of considering geometry effects of rigid caisson foundations on the SSI modelling. A major bridge supported on rigid double D caissons located in a highly active seismic zone is chosen to bring out the importance of including the geometry effects in case of deep foundations. The obtained results in terms of stiffness values are used to estimate the fundamental frequency of the bridge. The same is compared with the frequencies obtained using other available approaches. Furthermore, a final check of fundamental frequency of the bridge pier system is performed with 3D finite element analysis on PLAXIS 3D.

2 Extraction of Lumped Spring Parameters from 3d Fea

A method has been described by Jalbi et al. (2017) to compute the three stiffness terms (K_L, K_{LR}, and K_R) from advanced three dimensional Finite Element Analysis (FEA). A schematic view of the loading conditions along with the idealization is shown in Fig. 2. The vertical stiffness (K_V) is not expected to play a significant role as the caisson foundations are vertically stable and the influencing wave forms are horizontal shear waves. Hence, the K_V is neglected in this study. The linear range of a load-deformation curves can be used to estimate foundation rotations and deflections based on Eq. 1.

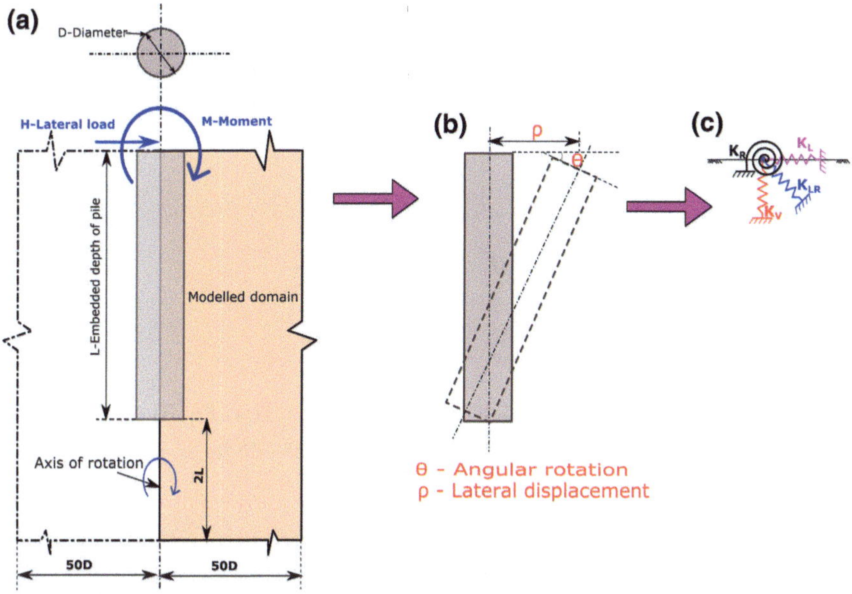

Fig. 2. Schematic representation of **a** soil domain with pile **b** rigid body rotation and translation and **c** idealization as lumped springs

$$\begin{bmatrix} H \\ M \end{bmatrix} = \begin{bmatrix} K_L & K_{LR} \\ K_{LR} & K_R \end{bmatrix} \begin{bmatrix} \rho \\ \theta \end{bmatrix} \tag{1}$$

Equation 1 can be re-written as Eq. 2 where I (Flexibility Matrix) is a 2 × 2 matrix given by Eq. 3

$$\begin{bmatrix} \rho \\ \theta \end{bmatrix} = [I] \times \begin{bmatrix} H \\ M \end{bmatrix} \tag{2}$$

$$I = \begin{bmatrix} I_L & I_{LR} \\ I_{RL} & I_R \end{bmatrix} \tag{3}$$

To obtain the equation unknowns, one can run a numerical model for a lateral load (say H = H_1) with zero moment (M = 0) and obtain values of deflection and rotation (ρ_1 and θ_1). The results can be expressed through Eqs. 4–5.

$$\begin{bmatrix} \rho_1 \\ \theta_1 \end{bmatrix} = \begin{bmatrix} I_L & I_{LR} \\ I_{RL} & I_R \end{bmatrix} \times \begin{bmatrix} H_1 \\ 0 \end{bmatrix} \tag{4}$$

$$\rho_1 = H_1 \times I_L \Rightarrow I_L = \frac{\rho_1}{H_1} \tag{5}$$

$$\theta_1 = H_1 \times I_{RL} \Rightarrow I_{RL} = \frac{\theta_1}{H_1} \tag{6}$$

Similarly, another numerical analysis can be done for a defined moment (M = M_1) and zero lateral load (H = 0) and the results are shown in Eqs. 7–8.

$$\begin{bmatrix} \rho_2 \\ \theta_2 \end{bmatrix} = \begin{bmatrix} I_L & I_{LR} \\ I_{RL} & I_R \end{bmatrix} \times \begin{bmatrix} 0 \\ M_1 \end{bmatrix} \tag{7}$$

$$\begin{aligned} \rho_2 &= M_1 \times I_{LR} \Rightarrow I_{LR} = \frac{\rho_2}{M_1} \\ \theta_2 &= M_1 \times I_R \Rightarrow I_R = \frac{\theta_2}{M_1} \end{aligned} \tag{8}$$

From the above analysis (Eqs. 4–8), terms for the I matrix (Eq. 3) can be obtained. Equation 2 can be rewritten as Eq. 8 through matrix operation.

$$[I]^{-1} \times \begin{bmatrix} \rho \\ \theta \end{bmatrix} = \begin{bmatrix} H \\ M \end{bmatrix} \tag{9}$$

Comparing Eqs. (1 and 9), the relation between the stiffness matrix and the inverse of flexibility matrix (I) given by Eq. 10. Equation 11 is a matrix operation which can be carried out easily to obtain K_L, K_R and K_{LR}.

$$K = \begin{bmatrix} K_L & K_{LR} \\ K_{RL} & K_R \end{bmatrix} = I^{-1} = \begin{bmatrix} I_L & I_{LR} \\ I_{RL} & I_R \end{bmatrix}^{-1} \tag{10}$$

$$K = I^{-1} = \begin{bmatrix} \frac{\rho_1}{H_1} & \frac{\rho_2}{M_2} \\ \frac{\theta_1}{H_1} & \frac{\theta_2}{M_2} \end{bmatrix}^{-1} \tag{11}$$

Therefore, mathematically, two FEA analyses are required to obtain the three spring stiffness terms. It is important to note that the above methodology is only applicable in the linear range and it is advisable only to use the obtained stiffness values for Eigen frequency analysis or a first estimate of the deformations using Eq. 1.

3 Numerical Model

Finite element analysis package PLAXIS 3D has been used in this study where the soil is idealized as an isotropic linear elastic material. A "Rigid Body" option has been set to the foundation where it is restricted to deform axially or in bending, and only the surrounding soil is mobilized. This assumption is valid since the well foundation has a low aspect ratio (due to the high diameter/width) and also because concrete has higher flexural and shear stiffness than soil. The interface between the soil and foundation had the same stiffness properties as the surrounding soil and a very fine mesh was implemented for enhanced accuracy.

The extent of the soil contour was taken as at least 50D (D is the diameter) and the depth h (h is the depth of the soil stratum) was at least twice that of the foundation. The objective was to ensure the stresses in the soil are not affected by the vicinity of the translational boundary conditions at the sides and bottom face (Figs. 2 and 3). Previous work presented in Krishnaveni et al. (2016) modelled the stratum with 5D width, whilst

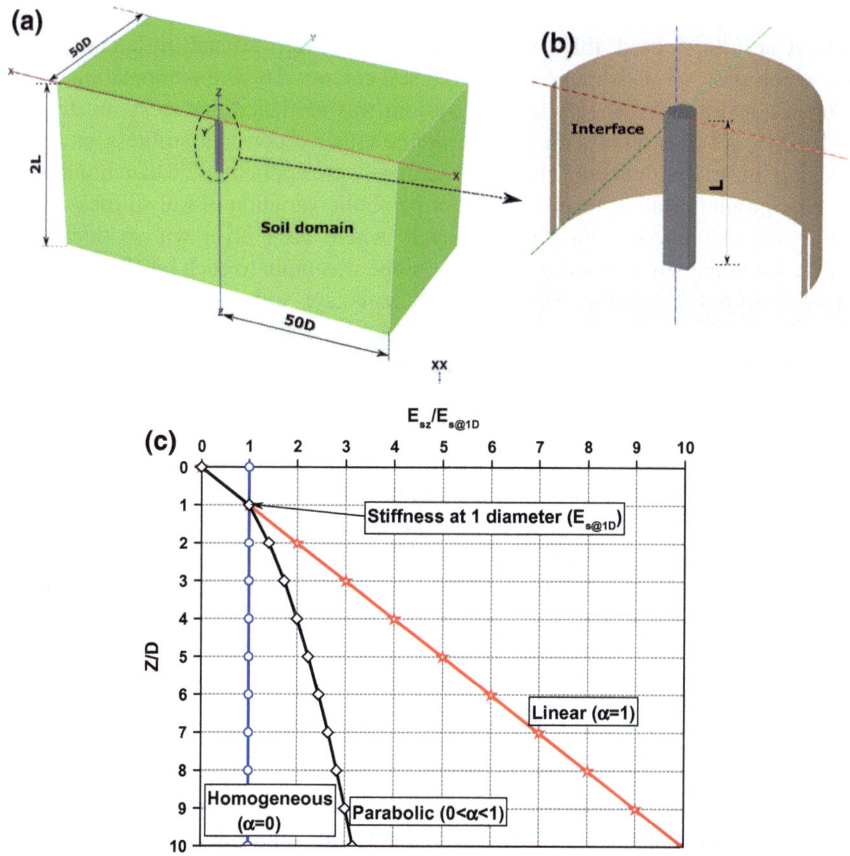

Fig. 3. **a** Numerical model geometry **b** Double D with the rigid SSI **c** Stiffness variation with depth

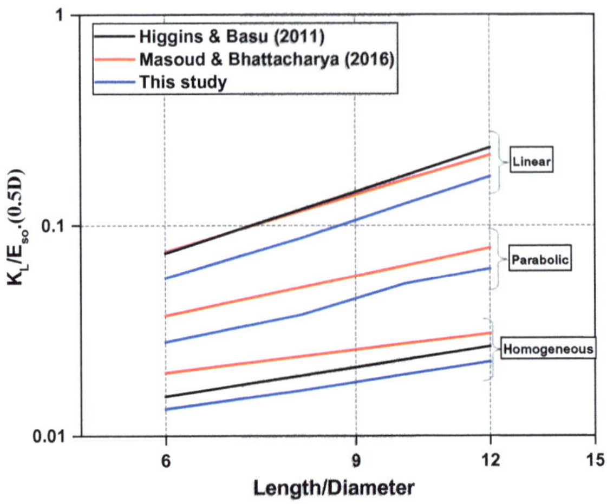

Fig. 4. Variation of normalized lateral stiffness

Abbas et al. (2008) used 40D on finite element software to analyze laterally loaded piles, which shows a wide range of selected soil extents. To save computational power and operational time cost, only half the system was modelled due to symmetry.

PLAXIS 3D also allows the user to define either a constant stiffness or stiffness increasing linearly with depth. These two settings were applied to idealize homogenous and linear ground profiles, respectively. For parabolic variation of soil stiffness, the soil stratum was discretized with multiple layers where each layer with a thickness of 0.025 h. An initial stiffness value and linear slope was input to each layer to represent a parabolic stiffness variation. Homogeneous soils are soils which have a constant stiffness with depth such as over-consolidated clays. On the other hand, a linear profile is typical for normally consolidated clays [or "Gibson Soil" (Gibson 1974)] and parabolic behaviour can be used for sandy soils, see Fig. 3c.

The software also has the capability to model the initial stresses in the stratum and the change in the stress state due to the construction sequence. Accordingly, the displacements were set to zero prior to application of the loads and values for K_L, K_R, and K_{LR} were computed. The first phase of loading consists of a lateral load of 100 kN (X-axis) with no moments acting while the second phase consists of 100 kN moment in the secondary axis (Y-axis) with no lateral loads. The displacements (both lateral and axial) are monitored during the loading and the stiffness values are evaluated as explained earlier.

3.1 Model Validation

In order to check the efficiency of the numerical modelling and the extracted results, the developed model is validated with the published literature (Higgins and Basu 2011; Shadlou and Bhattacharya 2016) of rigid circular foundations as there are no solutions for double D foundations. Higgins and Basu (2011) proposed non-dimensional formulations to estimate the stiffness of rigid piles in soils of uniform stiffness (homogeneous) and linearly varying stiffness. Similarly, Shadlou and Bhattacharya (2016) proposed such stiffness formulations for both rigid and flexible piles for a wide range of L/D ratio, where L represents the embedded depth and D is the diameter of the pile.

To validate the developed model results for double D, a circular shaft of diameter 5 m is chosen with varying length (30–60 m) representing L/D ratio of 6–12. The soil column is modelled using the three profiles, homogeneous, linear and parabolic stiffness variation. The elastic modulus of the soil is chosen as 100 MPa for the homogeneous profile while for linearly varying stiffness, the incremental stiffness is calculated based on the diameter of the shaft. However, for the parabolic stiffness, soil is discretized into many layers and then the corresponding initial stiffness and the incremental stiffness for each layer are defined accordingly. The loading is applied as discussed earlier.

Once the lateral and axial displacements at the caisson head are monitored for each loading, then the rotation is calculated using which the stiffness values are estimated. The stiffness functions are normalized using the soil stiffness (E_{so}) and the radius of the section (r). Figure 4 presents the variation of normalized length (L/D) with the normalized lateral stiffness for different soil profiles along with the comparative literature. It can be inferred from the Fig. 4 that the present model shows a consistent trend with increase in L/D ratio in spite of the nominal difference in the magnitude. This difference is mainly attributed due to the numerical modelling and can be ignored as the magnitude of difference is ±5%. Similarly, normalized K_R and K_{LR} are also presented in Figs. 5 and 6. Hence, the method of extraction is appropriate and can be used to estimate the stiffness of double D shaped foundations.

Once the stiffness of the foundation is established, the results can be used for the quick dynamic analysis of the structures supported on such foundations. Such dynamic analysis was performed by Arany et al. (2016) and Shadlou and Bhattacharya (2016) to identify the fundamental frequency of offshore wind turbines on using the lumped soil springs.

3.2 Saraighat Bridge

A bridge (Saraighat Bridge) supported on Double D caisson foundation is chosen to illustrate the significance of considering the geometry in the analysis. Saraighat Bridge is a double decker bridge supporting both roadway and railway over the piers and resting on double D caissons, see Fig. 7. The schematic representation of the bridge with the foundation details is shown in Fig. 8. This bridge is located in a very high seismic active zone in India (Assam). Further details about the bridge can be found in Dammala et al. (2017a).

The double D shaped caisson in this case is represented by two parameters for easy presentation, the diameter D and the width B as shown in Fig. 8c. It is modelled as (Fig. 8d) a rigid body due to the sheer size of the foundation. The embedment depth of the foundation is 41 meters and the surrounding soil is modelled using the homogeneous, linear and parabolic variation of stiffness with the appropriate soil parameters considered from Dammala et al. (2017b). Positive interface elements are considered with the stiffness of interface equals to that of surrounding soil (Fig. 8d). A horizontal

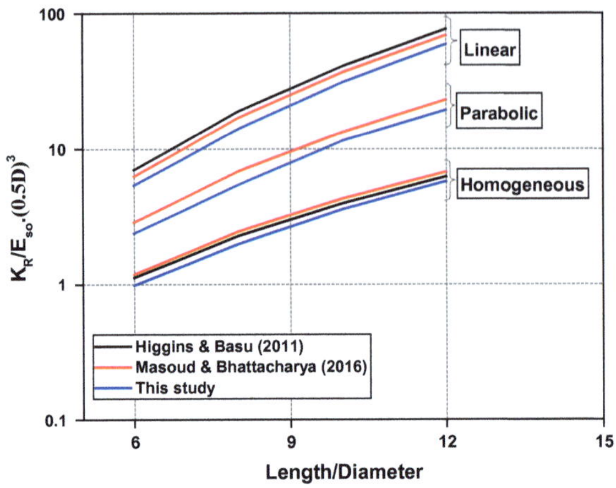

Fig. 5. Variation of normalized rocking stiffness

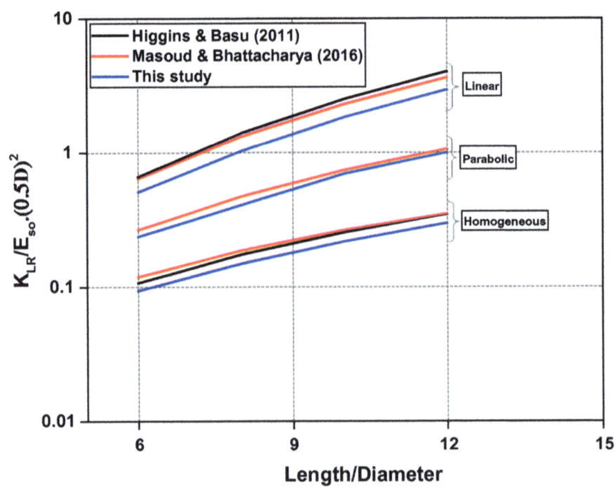

Fig. 6. Variation of normalized coupled rocking and lateral stiffness

load of 100 kN and a moment of 100 kN m is applied at the ground surface (bed level). Firstly, the extracted K_L, K_R, and K_{LR} values obtained directly from the FEA will be compared to the representative circular shaft computed based on solutions provided in the literature.

4 Results

Initially, two types of analysis is performed in homogeneous soil conditions, one with the double D shaped and the other with the circular shaft of representative diameter (9.6 m). Table 1 lists the horizontal (ρ_x) and axial (ρ_z) deformations at shaft head, for

Fig. 7. Saraighat Bridge, Assam, India

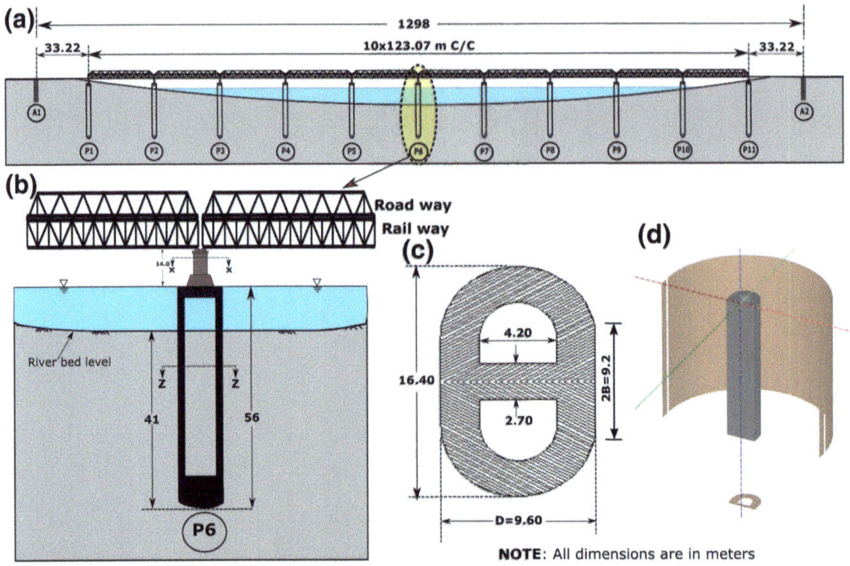

Fig. 8. Schematic view of **a** Saraighat Bridge, **b** central pier, **c** well foundation and **d** developed numerical model with interaction

both the cases in all the soil conditions. It is evident from the results that the displacements experienced by the double D caisson are significantly lower than that of representative circular shaft testifying the efficiency of double D over circular shafts.

Table 1. Displacements in lateral and axial directions in both the cases for homogeneous soil condition

Displacement direction	Circular section		Real geometry		% Difference	
	H = 100, M = 0	H = 0, M = 100	H = 100, M = 0	H = 0, M = 100	H = 100, M = 0	H = 0, M = 100
Lateral	0.08	0.002	0.05	0.007	60	185
Axial	0.01	0.004	0.001	0.0002	900	1900

The stiffness values obtained from the analyses for the double D are compared with the stiffness estimated for the circular shaft based on Shadlou and Bhattacharya (2016) rigid shaft formulations. Table 2a–c present the results in terms of stiffness values (K_L, K_R and K_{LR}) for both the geometrical conditions, at each soil profile. It is quite interesting to note that the stiffness of double D caisson is remarkably higher than that of circular shafts. An increase of almost 30–50% is noted in all the cases. This increase in impedance is obviously attributed due to the higher resistance offered by the rectangular portion of the double D caisson.

Table 2. Comparison of lateral stiffness obtained from circular assumption and real geometry

a

Profile	Lateral stiffness (K_L)		
	Circular (Shadlou and Bhattacharya 2016)	Real geometry (PLAXIS)	%Δ
Homogeneous	7.66	9.77	27
Linear	20.63	29.24	41
Parabolic	12.13	17.07	40

b

Profile	Rotational stiffness (K_R)		
	Circular (Shadlou and Bhattacharya 2016)	Real geometry (PLAXIS)	%Δ
Homogeneous	5328	7594	42
Linear	19890	30358	52
Parabolic	10712	15156	41

c

Profile	Coupled stiffness (K_{LR})		
	Circular (Shadlou and Bhattacharya 2016)	Real geometry (PLAXIS)	%Δ
Homogeneous	156	204	30
Linear	597	869	45
Parabolic	296	419	41

5 Natural Frequency Computation—Application

The simulated values of K_L, K_R and K_{LR} of the double D caisson are utilized in estimating the fundamental frequency (f_{nz}) of the bridge piers. As the underlying soil near the bridge is predominantly sandy soil with varying density along the depth (Dammala et al. 2017b), a linear variation of stiffness assumption would suit better, and hence adopted the stiffness functions of linear stiffness variation from the present study. A simple stick model (Fig. 1c) with a lumped mass at the pier head is considered to model the bridge pier system with the properties presented in Dammala et al. (2017a). A f_{nz} of 1.05 Hz is obtained by performing linear Eigen vector modal analysis on the soil well pier system. Debnath et al. (2016) estimated the f_{nz} of the Saraighat Bridge pier system to vary between 0.9031–0.9119 Hz based on operational modal analysis. Apart from this, Dammala et al. (2017a) performed similar Eigen vector modal analysis using the distributed spring approach and arrived at an f_{nz} of 0.8547 Hz.

In order to better check the present results, a three dimensional (3D) finite element program (PLAXIS 3D) is employed to model the entire support system of the bridge at the central part (Pier 6). The caisson is embedded in the loose sandy soil from the ground surface to a depth of 11 m followed by a dense deep sand of 30 m thickness, further followed by a clay layer of 30 m deep. Appropriate structure and soil properties along with the loading on the bridge deck are considered from the literature (Dammala et al. 2017a). Figure 9 shows the developed model. A static lateral load is applied at the pier head to perform a free vibration analysis of the structure (Fig. 9b). The load is taken off in the next phase allowing the structure to vibrate freely (50 s) in order to monitor the free decay of the system. Similar analysis is presented for a building in the tutorial manual of PLAXIS 3D (PLAXIS 2013).

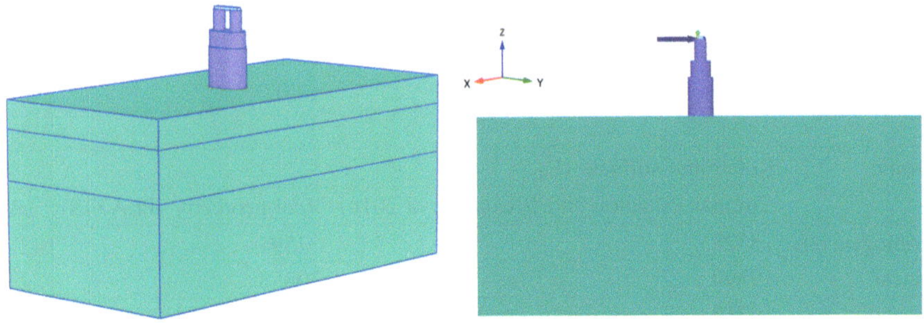

Fig. 9. Numerical model of the entire Soil-Well-Pier system along with the pier head loading

The deformed contours of the soil well pier system is shown in Fig. 10, where almost no significant displacements can be observed in the soil while the pier experienced a maximum displacement of 0.557 mm. The free vibration decay of the system is shown in Fig. 11a, b. The enlarged view of the free decay as shown in Fig. 11b represents a fundamental period of 1.30 s (f_{nz} = 0.769 Hz). Although this is narrowly different to the monitored frequency of the bridge (0.90 Hz), it gives an idea that the range of frequency could possibly fall between 0.70 and 1 Hz. The f_{nz}

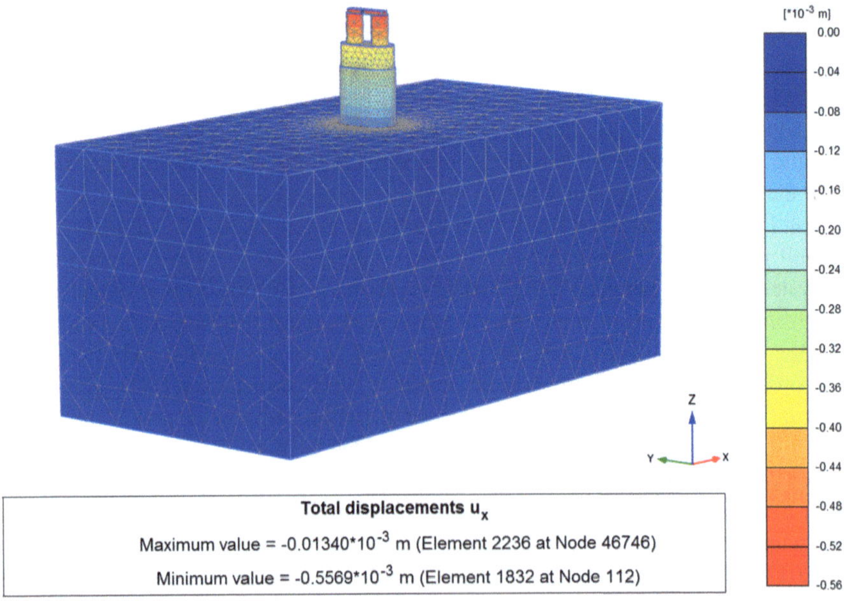

Fig. 10. Deformation contours of the soil well pier system

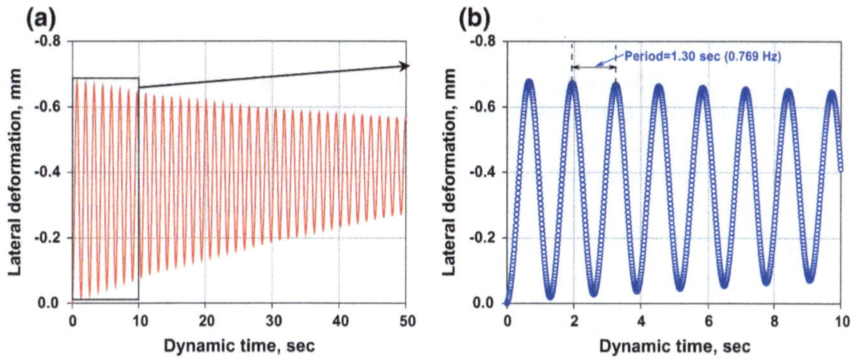

Fig. 11. **a** Free vibration decay of the SWP system, **b** Enlarged view briefing the fundamental period

obtained from the lumped spring approach could satisfactorily catch the response of the system although the stiffness functions utilized are obtained from the static impedance functions without considering the soil stiffness degradation and damping characteristics. It may well be inferred that on inclusion of stiffness degradation, the f_{nz} may further decrease which could possibly yield a closer agreement with the monitored data. Table 3 lists the f_{nz} from various approaches and the percentage difference.

Table 3. Comparison of f_{nz} using different approaches

Approach	Monitored data (Debnath et al. 2016)	Distributed springs (Dammala et al. 2017a)	Lumped springs (This study)	Numerical approach (PLAXIS 3D)
f_{nz} (Hz)	0.903	0.855	1.05	0.769
%Δ with monitored data	–	−5.3	+16.28	−14.84

6 Conclusions

The available stiffness formulations for the analysis of rigid deep foundations are based on either traditional circular shafts or on rectangular or square geometry. This study highlights the significance of having the geometry based non-dimensional formulations for stiffness estimation with a case study. A bridge located in a highly active seismic zone resting on massive double D caisson shaped foundations was chosen for illustration. The stiffness of the caisson in both the cases (circular and double D) is estimated using advanced finite element program. Three soil profiles with homogeneous, linear and parabolic stiffness variation are considered in the analysis. The results indicate that the geometry of the foundation do effect the stiffness, mostly the rotational stiffness due to the resistance offered by the rectangular section. Furthermore, the developed stiffness values are utilized in determining the fundamental frequency of the bridge pier system and a reasonably close match is achieved with the extracted stiffness values. Further soil stiffness degradation and damping characteristics need to be considered to enhance the efficiency of the highlighted geometry effects. Since the fundamental frequency calculated using the lumped springs provided a reasonable estimate, further scope of the work is required to develop such geometry based stiffness non-dimensional forms for easy-to-use in the design offices for preliminary analysis and design.

Acknowledgements. The first author would like to thank the Commonwealth Scholarship Commission (CSC) for providing him the financial support (reference no: INCN-2016-214) enabling him to perform his research activities in the University of Surrey, United Kingdom. The funding received is fully acknowledged.

References

Abbas, J.M., Chik, Z.H., Taha, M.R.: Single pile simulation and analysis subjected to lateral load. Electron. J. Geotech. Eng. **13** (2008)

Arany, L., et al.: Closed form solution of eigen frequency of monopile supported offshore wind turbines in deeper waters incorporating stiffness of substructure and SSI. J. Soil Dyn. Earthq. Eng., Elsevier, (2016). https://doi.org/10.1016/j.soildyn.2015.12.011

Banerjee, P.K., Davies, T.G.: The behavior of axially and laterally loaded single piles embedded in non-homogeneous soils. Geotechnique (1978). https://doi.org/10.1680/geot.1978.28.3.309

Carter, J.P., Kulhawy, F.H.: Analysis of laterally loaded shafts in rock. J. Geotech. Eng. ASCE (1992). https://doi.org/10.1061/(ASCE)0733-9410(1992)118:6(839)

Dammala, et al.: Scenario based seismic re-qualification of caisson supported major bridges—a case study of Saraighat Bridge. J. Soil Dyn. Earthq. Eng. (2017a). https://doi.org/10.1016/j.soildyn.2017.06.005

Dammala, et al.: Dynamic soil properties for seismic ground response studies in Northeastern India. Soil Dyn. Earthq. Eng. **100**, 357–370 (2017b). https://doi.org/10.1016/j.soildyn.2017.06.003

Debnath, N., et al.: Multi-modal vibration control of truss bridges with tuned mass dampers under general loading. J. Vibr. Control (2016). https://doi.org/10.1177/1077546315571172

Gazetas, G.: Formulas and charts for impedances of surface and embedded foundations. J. Geotech. Eng. ASCE (1991). https://doi.org/10.1061/(ASCE)0733-9410(1991)117:9(1363)

Gibson, R.: The analytical method in soil mechanics. Geotechnique **24**, 115–140 (1974)

Higgins, W., Basu, D.: Fourier finite element analysis of laterally loaded piles in elastic media. International Geotechnical Report, University of Connecticut, US (2011)

Jalbi et. al.: Practical method to estimate foundation stiffness for design of offshore wind turbines. Wind Energy Eng.: A handbook for onshore and offshore wind turbines, 329 (2017)

Krishnaveni, B., Alluri, S.K.R., Murthy, M.R.: Generation of p-y curves for large diameter monopiles through numerical modelling. Int. J. Res. Eng. Technol. (2016)

Plaxis, B.V.: PLAXIS 3D 2013 Reference Manual. PLAXIS BV, Delft (2013)

Poulos, H.: The displacement of laterally loaded piles: I-single piles. J. Soil Mech. Found. Div. **97**, 711–731 (1971)

Shadlou, M., Bhattacharya, S.: Dynamic stiffness of monopoles supporting offshore wind turbine generators. J. Soil Dyn. Earthq. Eng., Elsevier, (2016). https://doi.org/10.1016/j.soildyn.2016.04.002

Modelling of Hydro-Mechanical Coupling in Land Uplift Due to Groundwater Recharge

Pei-Hsun Tsai(✉), Jian-Han Lin, and Syuan-Yi Wang

Department of Construction Engineering, Chaoyang University of Technology, Taichung, Taiwan
phtsai@cyut.edu.tw

Abstract. Artificial groundwater recharge is used to increase groundwater in areas with water scarcity. After groundwater recharge, the water table rises, causing ground surface uplift from the increased pore water pressure in an aquifer. To manage groundwater resources effectively, understanding the hydro-mechanical features of aquifers during groundwater withdrawal and recharge is necessary. In practice, an artificial pool can be used to collect surface runoff so that it infiltrates the aquifer and recharges groundwater. Additional studies are required to understand the effect of various parameters on groundwater recharge and ground surface uplift. This study used the finite difference software FLAC 8.0 to examine the influence of the initial degree of saturation and soil type on the rate of groundwater recharge and ground surface uplift in an aquifer. Each aquifer was in an unsaturated state before groundwater recharging; therefore, groundwater recharge analyses were simulated by the two-phase flow in an unsaturated porous media. The results showed that the ground surface uplift was the largest when the initial degree of saturation was 70%, sequentially followed by a saturation of 60 and 50%. The influence of the initial degree of saturation was negligible on the cumulative groundwater recharge volume during the study period. Regarding the influence of soil types on ground surface uplift and the cumulative groundwater recharge volume, the results indicated that the sandy aquifer had the largest cumulative groundwater recharge volume during the study period. However, the ground surface uplift in the sandy aquifer was minimal.

1 Introduction

In recent years, water requirement has increased because of industrial development and population growth. Excessive groundwater withdrawal from aquifers causes groundwater depletion, seawater intrusion, groundwater salinization, and ground settlement. Groundwater recharge is performed to increase groundwater resources. After groundwater recharge, the water table rises resulting in ground surface uplift caused by increased pore water pressure within an aquifer. Although several studies have examined land settlement caused by groundwater withdrawal, only a few have investigated ground surface uplift caused by groundwater recharge, potentially because small-scale ground surface uplift does not cause environmental hazards or damage infrastructure. To manage groundwater resources efficiently, the hydro-mechanical

S. Wang et al. (eds.), *New Solutions for Challenges in Applications of New Materials and Geotechnical Issues*, Sustainable Civil Infrastructures,
https://doi.org/10.1007/978-3-319-95744-9_6

features of aquifers during groundwater withdrawal and recharge must be identified. In practice, an artificial pool can be used to collect surface runoff so that it infiltrates an aquifer and recharges groundwater. The artificial pool may be used as a stormwater detention pool to reduce flood disasters and recharge groundwater.

Various studies have recently examined ground surface uplift caused by groundwater recharge. Giao et al. (1999) used a finite element method code to calculate the compression and rebound of a multi-aquifer in artificial recharge. Bawden et al. (2001) used global positioning system data with interferometric synthetic aperture radar (InSAR) imagery to examine groundwater withdrawal and recharge and found that it caused long-term subsidence of 12 mm/year. Bell et al. (2008) used the Permanent Scatterer InSAR (PSInSAR) methodology along with data on water level changes to investigate the temporal and spatial patterns of an aquifer in response to groundwater pumping and recharge in the Las Vegas Valley. Teatini et al. (2011) and Gambolati and Teatini (2015) systematically listed the studies related to ground deformation caused by the withdrawal and injection of fluid into geological formations.

Most studies on groundwater recharge have examined the influence of geological conditions and the dimensions of recharge pools on the volume of water that can penetrate an aquifer. Few hydro-mechanical coupling studies have examined ground surface uplift caused by groundwater recharge. The deformation caused by the groundwater recharge of unsaturated soil is more complex because the deformation of an aquifer is associated with effective stress distribution, pore water pressure, and saturation degree of the aquifer. The transient distribution of the degree of saturation during groundwater recharge should be calculated using a seepage analysis of unsaturated soil. The location of the groundwater recharge pool can also be selected after identifying the effect of influencing parameters on ground surface uplift. Therefore, additional studies are required to understand how various parameters affect groundwater recharge.

This study aimed to investigate ground surface uplift caused by groundwater recharge during one rainy season (90 days). Ground surface uplift depends on the material properties of an aquifer and the hydrogeological conditions. Therefore, numerical simulations were used to study the initial degree of saturation and soil type of the aquifer and understand how these influenced ground surface uplift and the cumulative groundwater recharge volume.

2 Two-Phase Flow Model

A two-phase flow model in finite difference software FLAC 8.0 (Itasca Consulting Group 2016) allows numerical modeling of the flow of water and gas through porous media. In the two-phase flow model, void space in porous media is filled with water and gas, which makes the model suitable for simulating and solving the problem of hydro-mechanical coupling in unsaturated soils. The pressure difference between gas and water is called capillary pressure, which is a function of the degree of saturation and can be represented by the van Genuchten model. For conciseness, only the water-related details (but not those related to gas) of the two-phase flow model are presented here. In unsaturated soil, the water transport is described using Darcy's law:

$$q_i = -k_{ij} k_r \frac{\partial}{\partial x_j}(P - \rho g_k x_k), \tag{1}$$

where the tensor k_{ij} is the saturated mobility coefficient, k_r is the relative permeability of water, P is pore pressure, ρ is density of water, and g is gravity acceleration. Relative permeability, k_r, is a function of the degree of saturation by the van Genuchten form (Itasca Consulting Group 2016):

$$k_r = S_e^b \left[1 - \left(1 - S_e^{1/a}\right)^a\right]^2, \tag{2}$$

where a and b are constant parameters and S_e is the effective saturation. In the van Genuchten model, parameter c was required to calculate the relative permeability of gas. The effective saturation is defined as

$$S_e = \frac{S - S_r}{1 - S_r}, \tag{3}$$

where S is the degree of saturation and S_r is the residual degree of saturation.

The capillary pressure P_c can be represented by the van Genuchten form:

$$P_c = P_g - P_w = P_o \left[S_e^{-1/a} - 1\right]^{1-a}, \tag{4}$$

where P_g is the pore pressure of gas, P_w is the pore pressure of water, and parameter P_o depends on the material properties. The constitutive law for water is:

$$n\left[\frac{S}{K_w}\frac{\partial P_w}{\partial t} + \frac{\partial S}{\partial t}\right] = -\left[\frac{\partial q_i}{\partial x_i}\right], \tag{5}$$

where n is porosity, K_w is the water bulk modulus.

In groundwater recharge, the degree of saturation of the unsaturated soil increases because of water infiltration into the soil. The capillary pressure and effective stress of the unsaturated soil decrease in groundwater recharge. The decreased effective stress causes ground surface uplift.

3 Parameter Estimate and Model Verification

In numerical analysis of this study, the two-phase flow model was adapted to perform the coupled fluid-mechanical calculation. Recharge model tests were performed to obtain the values for the two-phase flow model parameters required for numerical analysis and verify the rationality of the analysis. The parameter values of the two-phase flow model for three soils (sand, silt loam, and loam) were selected from the sandbox recharge model test developed by Liu (2001) and cylindrical specimen model tests performed in this study.

3.1 Sandbox Recharge Model Tests

A 10-cm-wide, 180-cm-long, 70-cm-high sandbox was used to conduct the tests by Liu (2001). The sandbox was filled with Ottawa standard sand [average particle size (D_{50}) = 0.35 mm and uniformity coefficient (c_u) = 1.8] layer-by-layer. A trapezoidal pit was then dug into the sand, with upper and lower base lengths of 30 and 10 cm, respectively, and height of 10 cm. The surface of the pit was covered with a layer of impermeable membrane. Water was added to the top of the membrane to fill the pit. Consistent with Pascal's principle, the water was connected to an external water source, so that the water level of the pit remained fixed in the experiments. At the beginning of the test, the impermeable membrane was gradually pulled away, and the wetting front of the sand was recorded by a digital camera. In the numerical simulations, the two-phase flow parameters of the sand were based on the soil-water characteristic curve obtained from Liu's experimental results (Liu 2001). A total of 2200 elements, each approximately 2 cm long, were discretized in a FLAC grid, as shown in Fig. 1. The mechanical boundary conditions corresponded to roller boundaries at the base and sides of the model. The boundaries on the bottom and right side of the model were assumed to be impermeable boundaries. The boundaries on the left side of the model and the surface of pit were assumed to be seepage boundaries. The non-wetting pore pressure was fixed above the water level. The wetting front (experimental results) and flow vector (numerical results) at 60, 600, and 900 s are shown in Fig. 2. The flow vectors (blue arrows in Fig. 2) were close to the wetting front. Therefore, the parameter values of the two-phase flow for the sand as well as the numerical analysis procedure were reasonable. Tables 1 and 2 list the parameter values of the two-phase flow model of the sand.

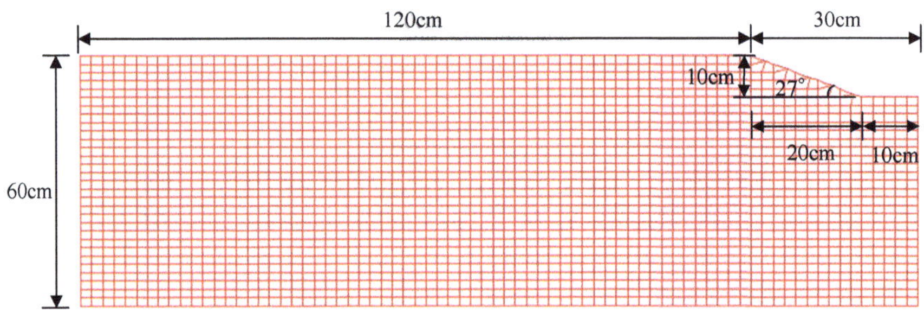

Fig. 1. The FLAC mesh of the sandbox recharge model test

3.2 Cylindrical Specimen Recharge Model Test

For the selection of parameter values of two-phase flow model for loam and silt loam, this study performed the cylindrical specimen recharge model tests and inverse analysis to estimate the optimal parameter values of the two-phase flow model for loam and silt loam soil. The loam (20% sand, 40% clay, and 40% silt) and silt loam (20% sand, 10% clay, and 70% silt) were separately placed into in an acrylic cylindrical mold (approximately 7 cm in diameter and 17 cm in height). The height of the specimens was 16 cm. An external water source was connected to the bottom of the mold so that

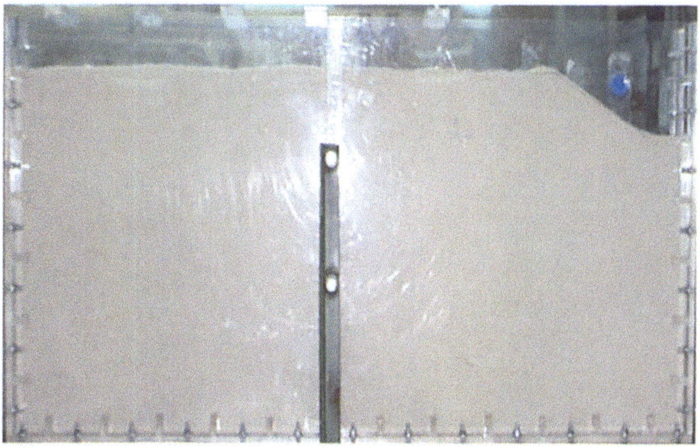

(a) at 0 s

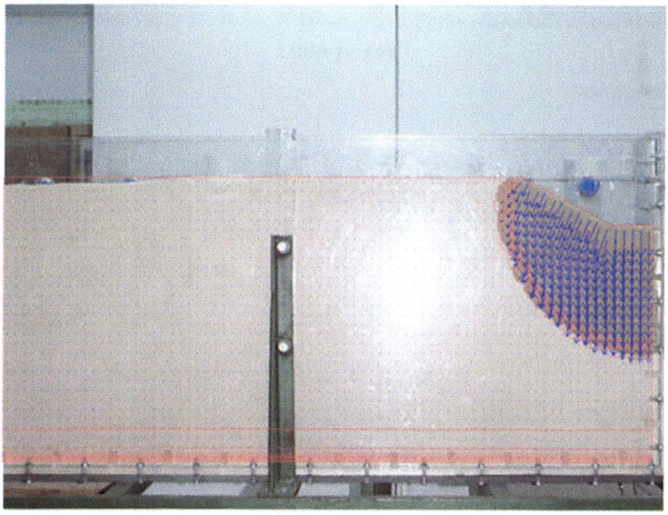

(b) at 60 s

Fig. 2. Comparison of numerical and experimental results in the sandbox recharge model test

the water level of the specimen could be maintained at a 12-cm height. The valve was opened at the beginning of the tests for the water penetrate the specimens from the bottom of the mold. A LVDT was placed on top of the mold to measure the uplift of the specimen. The numerical simulation was performed through axisymmetric analysis; a total of 5600 elements, each 0.1 cm long, were discretized in a FLAC grid, as shown in Fig. 3. The mechanical boundary conditions corresponded to roller boundaries at the base and sides of the model. The boundaries on the sides of the mesh were assumed to be impermeable boundaries. The non-wetting pore pressure was fixed on the top of the model; furthermore, a fixed wetting pore pressure was applied on the bottom of the model. The different parameter values in the two-phase flow model were substituted to

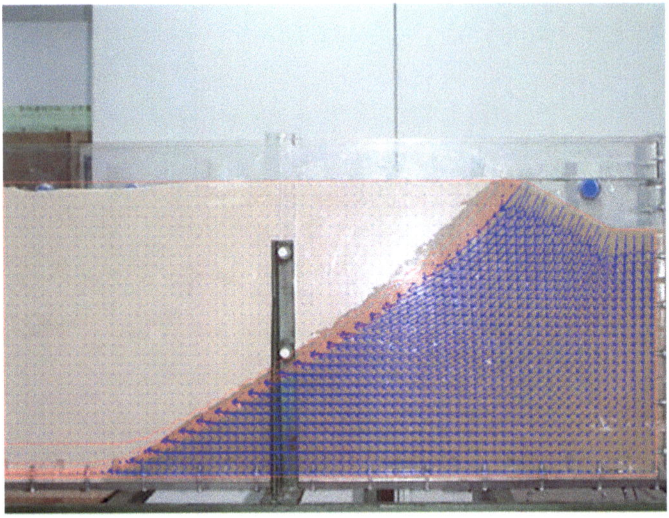

(c) at 600 s

(d) at 900s

Fig. 2. (*continued*)

calculate their corresponding numerical results. By comparing the experimental and numerical results, the optimal parameter values were obtained when the difference between the two results was minimal. The numerical simulation results for the optimal parameter values were compared with the experimental results, as presented in Fig. 4. Tables 1 and 2 list the parameter values of the two-phase flow model of the loam and silt loam soils.

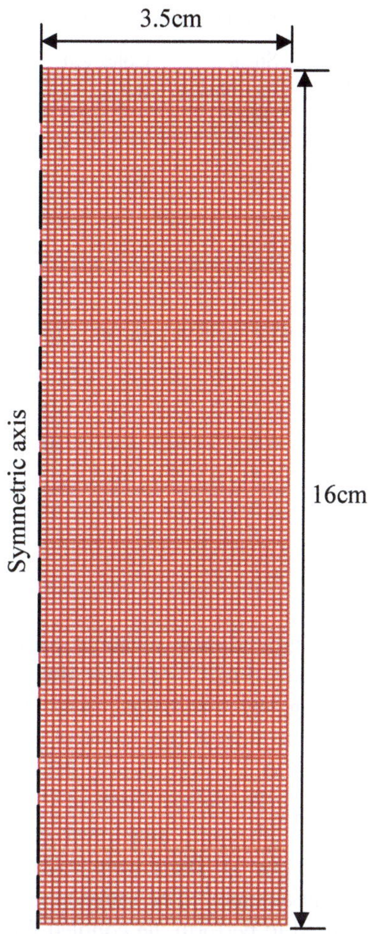

Fig. 3. The FLAC mesh of the cylindrical specimen recharge model test

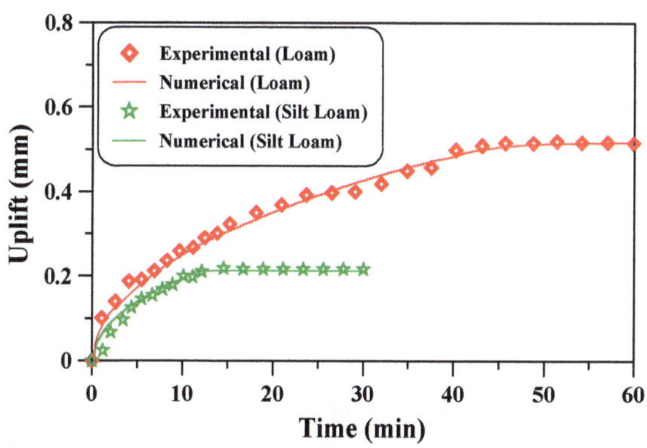

Fig. 4. Comparison of numerical and experimental results in the cylindrical specimen recharge model test

Table 1. The material parameters of the aquifers

Aquifer	Density (kg/m^3)	Bulk modulus (MP_a)	Shear modulus (MP_a)	Cohesion (kP_a)	Friction angle (degree)	Saturated permeability (m/s)
Sand	1800	16.6	7.67	–	38	4.8×10^{-4}
Silt loam	1850	12.7	3.91	3.1	26	3.5×10^{-5}
Loam	1826	11.9	3.34	5.2	23	3.7×10^{-6}

Table 2. The van Genuchten parameters of the aquifers

Aquifer	a	b	c	S_r	P_o (kP_a)
Sand	0.66	0.5	0.5	0.2	4.46
Silt loam	0.86	0.5	0.5	0.41	9.81
Loam	0.51	0.5	0.5	0.42	8.53

4 Numerical Model

To study the influence of the groundwater recharge parameters on ground surface uplift, a finite difference program, FLAC 8.0, was used to perform the numerical analyses of groundwater recharge by a recharge pool. The numerical model of the parametric analysis was similar to that of the sandbox recharge model test, and the aquifer model was rectangular, with a length, depth, and element size of 90 m, 20 m, and 0.5 m × 0.5 m, respectively. The initial depth of the water table was 8 m. The recharge pool was trapezoidal, with upper and lower base lengths of 10 and 4 m, respectively, and height of 3 m (Fig. 5). There were 3600 elements in the model. The material properties of soils were assumed to be satisfied to the Mohr-Coulomb model. Two-phase flow model was used to perform the coupled fluid-mechanical calculation. A symmetric numerical model was employed; the right-hand boundary of the model was a symmetrical axis.

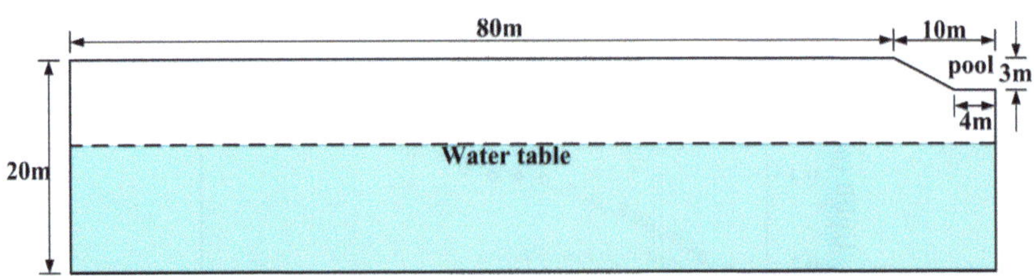

Fig. 5. Configuration of the groundwater recharge pool

The mechanical boundary conditions corresponded to roller boundaries at the base and sides of the model. Below the water table, the fluid-flow boundary condition was the fixed wetting pore pressure. The non-wetting pore pressure was fixed above the water level. The boundaries on the bottom and left side of the recharge pool were assumed to be seepage boundaries during the groundwater recharge period; furthermore, the non-wetting pore pressure was fixed during the empty pool period. The recharge pool was empty in the beginning. The water level raised to the maximum water level height (3 m) at a rate of 0.2 m/h. A seepage analysis of the groundwater recharge was calculated every time the water level rose by 1 m (three times). The mechanical analysis was performed to calculate the ground surface uplift after every seepage analysis. The time of a groundwater recharge cycle was assumed to be 360 days. The water level in the recharge pool was full for the first 90 days to recharge the groundwater. The groundwater recharge stopped on Day 90. The pool gradually became empty pool after groundwater recharge ceased. The evolution of ground surface uplift was continuously tracked until Day 360.

5 Parametrical Study

5.1 Influence of Various Initial Degrees of Saturation

Figure 6 shows the history curves of the cumulative groundwater recharge volume with different initial degrees of saturation of a sandy aquifer. The simulated groundwater recharge occurred for 90 days. It also illustrates that the cumulative groundwater recharge volume increased with time during the groundwater recharge period. The cumulative groundwater recharge volume remained unchanged when the groundwater recharged ceased. Furthermore, the history curves of the cumulative groundwater recharge volumes for different initial saturations of the sandy aquifer nearly coincided. The initial degree of saturation of the sandy soil had little effect on the history curve of the cumulative groundwater recharge volume.

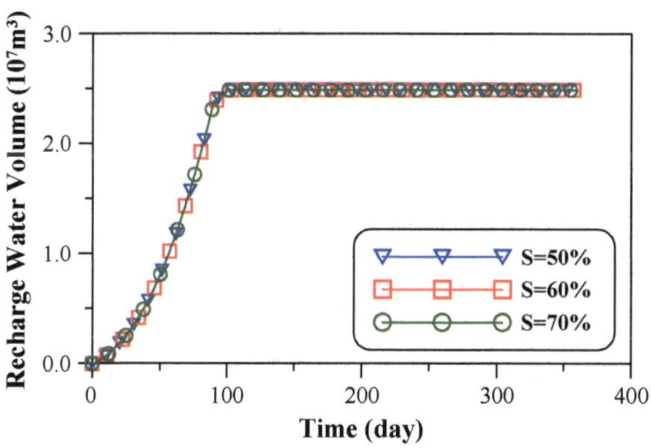

Fig. 6. Influence of initial degree of saturation on the cumulative groundwater recharge volume

Figure 7 illustrates the ground surface uplifts at different locations of the ground surface (the center of the recharge pool was the origin of the coordinate) on Day 90 for the groundwater recharge in the sandy aquifer with a different initial degree of saturation. The ground surface uplift at the end of the 90 days is reflected in the figure (the water level of the recharge pool was maintained at 3 m). The maximum ground surface uplifts on Day 90 at an initial degree of saturation of 50, 60, and 70% were 0.028, 0.03, and 0.031 m, respectively. Ground surface uplift within 40 m from the recharge pool was obvious; however, the ground surface uplift was relatively low beyond the recharge pool. The ground surface uplift increased slightly during the recharge period when the initial degree of saturation increased for the sandy aquifer.

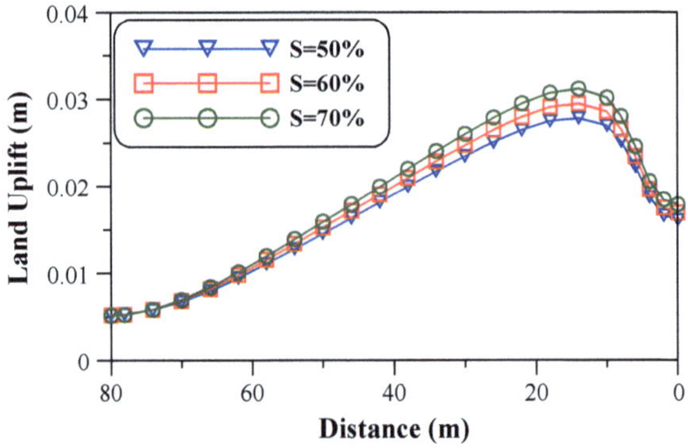

Fig. 7. Influence of initial degree of saturation on ground surface uplift (Day 90)

Figure 8 shows the ground surface uplift history curves at 14 m from the center of the recharge pool (x = 14 m) for different initial degree of saturations of the sandy aquifer during groundwater recharge. The ground surface uplift was the largest when the initial degree of saturation was 70%, sequentially followed by saturations of 60 and 50%. The ground surface uplift related to the water level of the recharge pool. When the water level of the recharge pool increased to the maximum level (3 m), the ground surface uplift also increased. When the water level of the recharge pool remained full, the ground surface uplift remained at a fixed value. Conversely, the ground surface uplift decreased as the water level of the recharge pool fell after the groundwater recharge period. It can be because that the left-side boundary of the mesh grid was assumed to allow water to flow out from the boundary freely. Therefore, when the groundwater recharge ceased, the ground surface gradually returned to its original position.

5.2 Influence of Various Soil Types

Figure 9 illustrates the history curves of the cumulative groundwater recharge volumes for different soil aquifers. The cumulative groundwater recharge volume of the sandy aquifer was the largest, followed by the silt loam aquifer. The loam aquifer had the lowest cumulative groundwater recharge volume, potentially because the cumulative groundwater recharge volume and saturated permeability coefficient of a soil aquifer are related. The permeability coefficient of unsaturated soil is related to the saturated permeability coefficient and the degree of saturation of the soil. When the permeability coefficient of the unsaturated soil aquifer is low, less water infiltrates the soil over a groundwater recharge period.

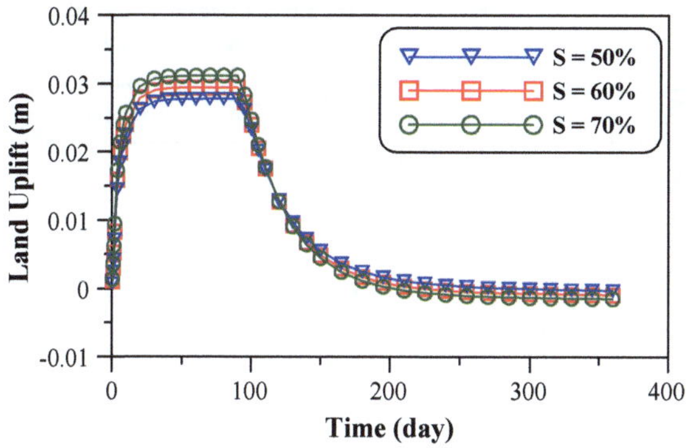

Fig. 8. Influence of initial degree of saturation on the history curve of ground surface uplift (at x = 14 m)

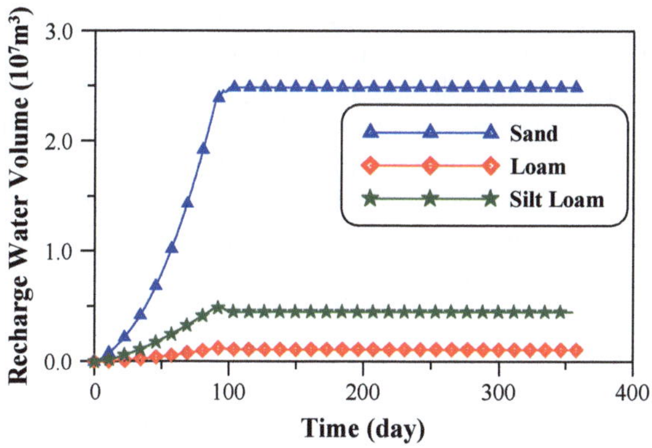

Fig. 9. Influence of soil type on the cumulative groundwater recharge volume

Figure 10 shows the ground surface uplift history curves at 14 m from the center of the recharge pool (x = 14 m) for different soil aquifers during the groundwater recharge period. The ground surface uplift of the silt loam aquifer increased with time during the groundwater recharge period. Ground surface uplift still increased in the silt loam aquifer even after the groundwater recharge period, but decreased after Day 105, probably because of the low permeability coefficient of silt loam. Therefore, groundwater could have been largely retained in the study area. Ground surface settlement occurred in the loam aquifer during the groundwater recharge period because the weight of the pool water was applied at the bottom of the pool. The settlement decreased over time because the pool water infiltrated the loam aquifer causing the soil to swell. The ground surface uplift was detected after Day 95. The lower permeability coefficient of the loam soil, because of its high clay content, resulted in less groundwater uplift in the loam aquifer. Moreover, during the groundwater recharge period, less water infiltrated the loam soil than it did the other soils because of lower permeability; in other words, the rate of ground surface uplift in the loam aquifer was relatively low.

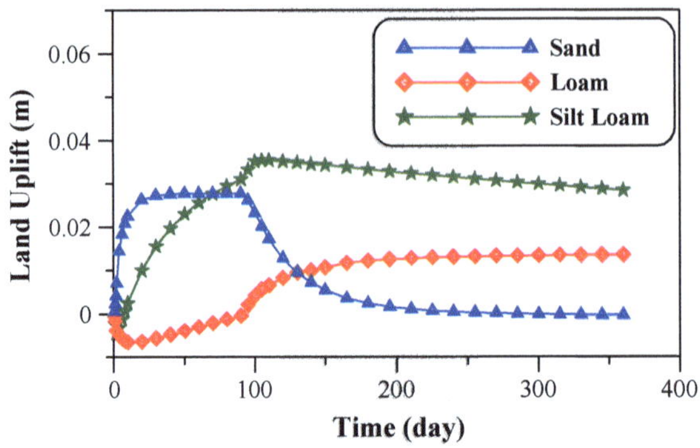

Fig. 10. Influence of soil type on the history curve of ground surface uplift (at x = 14 m)

Figure 11a, b demonstrates the relationship between the ground surface uplift and ground surface position at the end of the groundwater recharge period (Day 90) and that of the groundwater recharge cycle (Day 360), respectively. Figure 11a shows that the ground surface uplifts in the silt loam and sandy aquifers were obvious at the end of the groundwater recharge period (Day 90), particularly near the groundwater recharge pool. The maximum ground surface uplifts in the silt loam aquifer and sandy aquifer were approximately 3.1 and 2.8 cm, respectively. Figure 11b shows that the sandy aquifer had no obvious ground surface uplift at the end of the groundwater recharge cycle (Day 360), potentially because the groundwater flowed out of the study area because of the high permeability of sand. During this time, the groundwater in the silt loam aquifer flowed, but not beyond the study area, because of its low permeability. The zone of influence on ground surface uplift of the silt loam was wider than that of the other two soil types. On Day 360, the surface of the loam aquifer had an obvious uplift near the recharge pool. Other than the unloading effect caused by the empty recharge pool, the water that infiltrated the loam aquifer during the groundwater recharge period flowed to around, increasing the degree of saturation near the recharge pool, resulting in the obvious uplift.

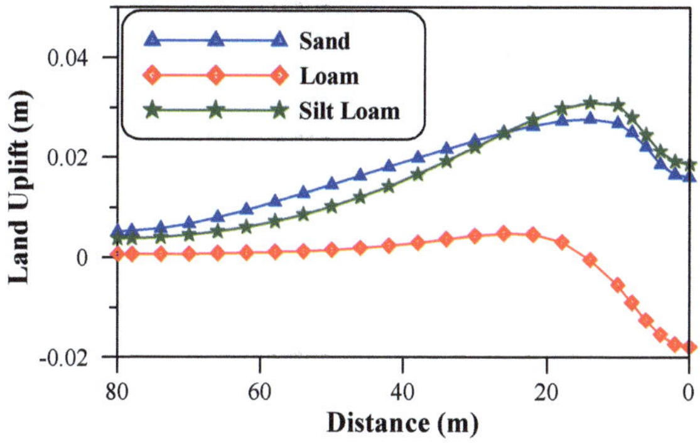

(a) Day 90

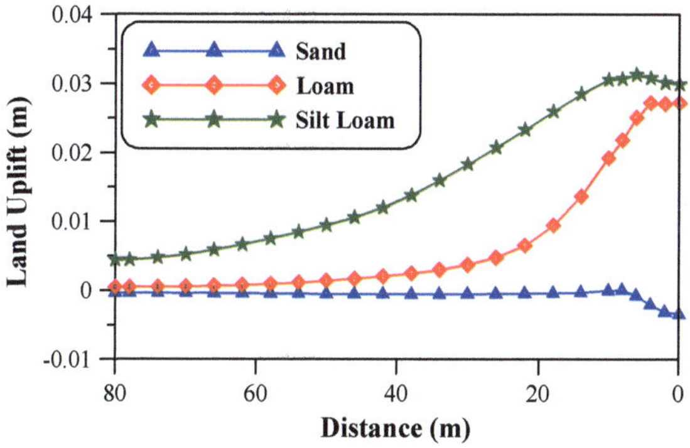

(b) Day 360

Fig. 11. Influence of soil type on ground surface uplift

6 Conclusions

By using the finite difference software FLAC 8.0 with a two-phase flow model, a parametrical study was examined ground surface uplift induced by groundwater recharge. The initial degree of saturation and different soil types on ground surface uplift and the cumulative groundwater recharge volume were discussed. The study results are summarized as follows:

1. Compared with the numerical and experimental results, it is clear that the numerical analysis used in this study is rational.
2. The initial degree of saturation had a non-significant effect on the cumulative groundwater recharge volume. When the initial degree of saturation of the aquifer was 70%, the ground surface uplift was the largest, sequentially followed by the initial degree of saturations of 60 and 50%.
3. The sand aquifer had the largest cumulative groundwater recharge volume, followed by the silt loam aquifer. The loam aquifer had the lowest cumulative groundwater recharge volume. At the end of the groundwater recharge period (Day 90), the ground surface uplifts in the silt loam and sandy aquifers were obvious. Because of

the high permeability of sand, the recharged groundwater flowed out of the study area. However, no obvious ground surface uplift in the sandy aquifer were noted at the end of the groundwater recharge cycle (Day 360). Constructing walls within a sandy aquifer to contain the groundwater may prevent the recharged groundwater flow out, it may be able to improve this situation.

Acknowledgements. This study was supported by the research funding from the National Science Council of Taiwan (NSC 102-2221-E-324-025-); their support is gratefully appreciated.

References

Bawden, G.W., Thatcher, W., Stein, R.S., Hudnut, K.W., Peltzer, G.: Tectonic contraction across Los Angeles after removal of groundwater pumping effects. Nature, Nature Publishing Group (2001). https://doi.org/10.1038/35090558

Bell, J.W., Amelung, F., Ferretti, A., Bianchi, M., Novali, F.: Permanent scatterer InSAR reveals seasonal and long-term aquifer-system response to groundwater pumping and artificial recharge. Water Resour. Res., Blackwell. (2008). https://doi.org/10.1029/2007WR006152

Gambolati, G., Teatini, P.: Geomechanics of subsurface water withdrawal and injection. Water Resources Research, Blackwell (2015). https://doi.org/10.1002/2014wr016841

Giao, P.H., Phien-Wej, N., Honjo, Y.: FEM quasi-3D modelling of responses to artificial recharge in the Bangkok multiaquifer system. Environ. Modell. Softw. Elsevier (1999). https://doi.org/10.1016/S1364-8152(98)00065-6

Itasca Consulting Group, Inc.: FLAC-fast lagrangian analysis of continua, Ver. 8.0 user's manual. Minneapolis, Minnesota, USA (2016)

Liu, C.J.: Influence of van Genuchten parameters on wetting front simulation and transit Seepage (in Chinese). Master thesis, Feng Chia University, Taichung (2001)

Teatini, P., Gambolati, G., Ferronato, M., Settari, A., Walters, D.: Land uplift due to subsurface fluid injection. J. Geodyn. Elsevier. (2011). https://doi.org/10.1016/j.jog.2010.06.001

Numerical Modelling of Reinforced Stone Columns and Bamboo Mattress for Supporting Causeway Embankment on Soft Soil Bed

Ardy Arsyad(✉), Lawalenna Samang, A. Bakri Muhiddin, Tri Harianto, and A. R. Djamaluddin

Department of Civil Engineering, Faculty of Engineering, Hasanuddin University, Makassar, Indonesia
ardyarsyad@yahoo.com

Abstract. This paper presents numerical model of reinforcement of causeway embankment over soft soil deposit using bamboo grid mattress and stone columns. A series of experimental tests were undertaken to obtain several mechanical parameters of stone columns, and mechanical characteristics of bamboo grid. The soft soil is silty clay in deep layer in which its index and engineering properties derived from oedometer tests. FEM model of a causeway embankment over bamboo grid mattress overlying deep soft soil reinforced by a group of granular columns encased with geotextile, was developed. To validate the FEM model, full scale experimental of similar model was conducted. It was found that FEM model is well agreement with the experimental model. The result explains the stress-strain behavior in bamboo grid mattress and stone columns, interacted with the soft soil as response to increasing embankment height. Bamboo grid mattress enhances the bearing capacity of the soft soil in supporting embankment leading to the decrease of settlements, while stone columns affect the acceleration of consolidation of the soft soil. The results would be beneficial for application of local natural materials such as bamboo for soft soil reinforcement as bamboo is widely available in developing Asian countries.

Keywords: Stone columns and bamboo mattress · Soft soil bed
Settlement · Drain · Reinforcement

1 Introduction

At infrastructure development in most countries, reliable infrastructures are required within high performance of stability and serviceability. Infrastructures have to be built in cost-effective foundation with sufficient load-bearing capacities and minimum long-term settlement. However, natural condition of lowland areas for infrastructure have become common problem since soft soil foundation in the area can cause excessive settlement which generates undrained failure of the infrastructure. Therefore, proper ground improvements are necessary to undertaken before constructing the infrastructure in order to prevent unacceptable excessive and differential settlement and increase bearing capacity of the foundations (Indraratna et al. 2013). Among other ground improvement techniques, stone column is one of world widely practised since the technique was

S. Wang et al. (eds.), *New Solutions for Challenges in Applications of New Materials and Geotechnical Issues*, Sustainable Civil Infrastructures,
https://doi.org/10.1007/978-3-319-95744-9_7

adopted in 1970's in the US. Stone columns can significantly improve bearing capacity, accelerate consolidation, increase slope stability, and control liquefaction (Fatahi et al. 2012). In comparison to other ground improvement such prefabricated vertical drains (PVDs), stone columns are stiffer and faster dissipation of excess pore water pressure from softclay (Basack and Rujiatkamjorn 2016; Guetif et al. 2007). The behaviour of stone column within typical fill embankment is that the total imposed stress can generate the increase of excess pore pressure, progressive settlement of soft clay and arching occurs due to the weight of fill arches over the stone columns (Low et al. 1994; Abusharar et al. 2009; Deb 2010). In such situation, stone columns are combined with geosynthetics reinforcement involving geocells and geogrids. In this study, geosynthetic materials can be alternatively replaced by an innovative and sustainable materials such as bamboo mattress as bamboo is cost effective, environmental friendly material, and posses higher tensile strength compared to geosynthetics (Hedge and Sitharam 2015). Recent study of the use of bamboo as soft soil reinforcement is bamboo cells (Hedge and Sitharam 2015). Even though, such issue of durability of bamboo in the natural condition, bamboo is able to resist towards such weathering in the fully saturated soil. The use of such ecomaterial of bamboo mattress as soft soil reinforcement combined with conventional stone columns remains endeavoured to be investigated. Therefore, this study examines the performance of embankment reinforced with stone columns and bamboo mattress, resting on soft soil bed with fully saturated.

2 Full Scale Experiment of Embankment Supported with Bamboo Mattress and Stone Columns

In order to examine the effect on stone columns with bamboo mattress on the performance of embankment on sofy soil bed, a full scale experiment was undertaken. In this way, embankment was constructed on the top of grid system of bamboo used as matrass, and granular columns underlying by soft soil beds (Fig. 1). The soft soil consists of a clay silt layer, underlying by silty clay, clay tuff, and gravel clay. The soft soil is around 4 meters deep, and stone columns were inserted, lying from the top (clay silt) to bottom of the layers (clay gravel). The stone columns are in 0.6 m diameter with 3 m length. Meanwhile, the size of the embankment is 10 m × 10 m × 3.5 m. The granular columns were encased with geotextile non-woven. Totally, stone columns are 16 columns with a 1.8 m space between each column (Fig. 2). The construction of embankment was in three stages, in which settlement measurement undertaken during the stage of embankment filling.

Laboratorium tests were conducted to obtain index and engineering properties of the soft soil, embankment, and stone columns (Table 1).

2.1 Measured Settlements of Reinforced and Unreinforced Embankments

Measurements of settlements were undertaken at the embankments within two different conditions such as unreinforced and reinforced embankment. At the unreinforced embankment, stage of construction was started by filling a 0.85 m embankment on the

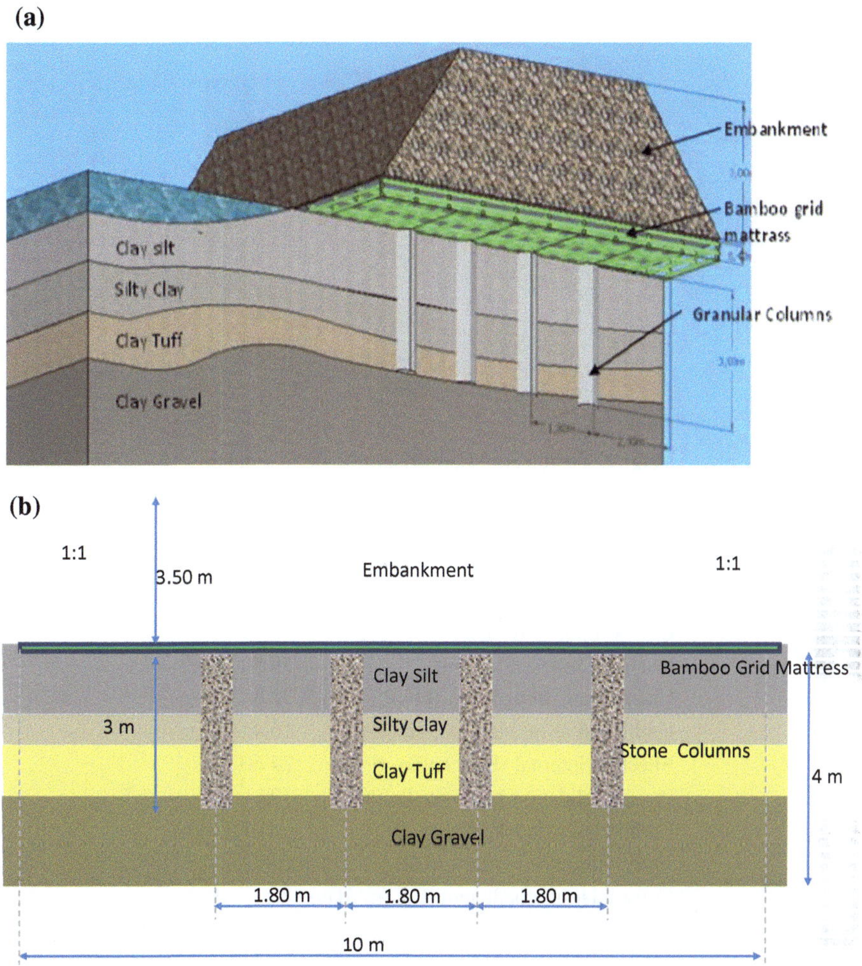

Fig. 1. Schematic (a) and cross sectional (b) views of reinforced embankment with bamboo mattress and stone columns.

ground, and the settlement was measured over 24 days. After that, in the second stage, additional 1.10 m thick embankment was laid on the top of the first stage of embankment, and the measurement of the settlement was undertaken for 18 days. In the last stage, 1.20 m embankment was laid on the top, and the observation of the settlement over 24 days (Fig. 3). It can be seen at Fig. 4 that primary settlement due to elastic response of the ground to embankment load was still happened until the end of construction. Total settlement was accounted for 75 cm.

At the reinforced embankment with bamboo grid mattress and granular columns, the first stage is filling 1.30 m embankment of the ground which was already set up with 16 granular columns and bamboo grid mattress. The observation of settlement was over 7 days. The second stage is 1.20 m thick embankment laid on the top of first embankment with a 50 days settlement observation (Fig. 3). At this stage, the consolidation settlement

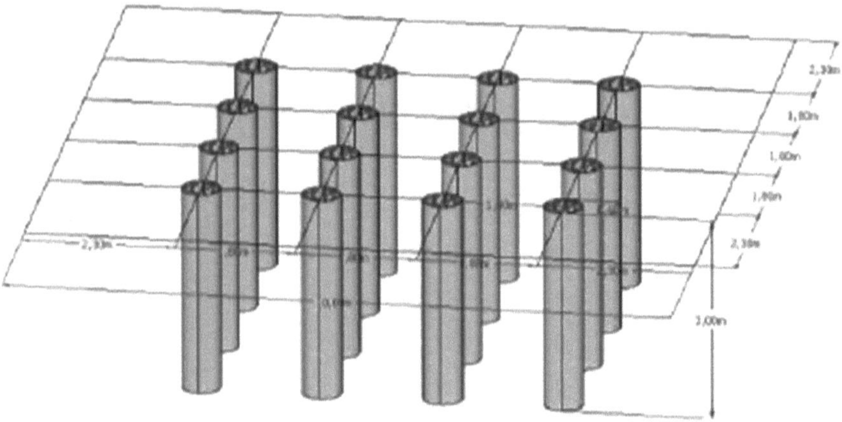

Fig. 2. Schematic 3-dimension of stone columns.

Table 1. Properties of the silty clay, embankment, and stone column.

Silty clay	
Unit weight, γ, (t/m^3)	1.689
Plasticity index, PI (%)	59.50
Unconfined strength, qu (kPa)	32.8
Undrained shear strength, cu (kPa)	16.4
Effective internal shear angle, ϕ' (°)	24.60
Pre-consolidation pressure, Pc (kPa)	103
Compression index, C_c	1.107
Void ratio, e_0	2.987
Consolidation coefficient, C_v, (m^2/day)	7.5×10^{-3}

Embankment	
Unit weight, γ, (t/m^3)	1.802
Cohesion, c (kPa)	0
Internal shear angle, ϕ, (°)	30

Gravel materials for stone column	
Specific gravity	2.90
Density, (kg/m^3)	2.88
Abrasion, (%)	23.36

was occurred and 90% consolidation can be achieved. The third stage of embankment is 1.10 m embankment with 50 days observation of settlement. Again, 90% consolidation was achieved at several days in the end of construction. Total settlement is 84 cm (Fig. 4). It is obvious that bamboo grid mattress and granular columns can accelerate settlement, and increase of the height of embankment that can be supported by soft soil ground.

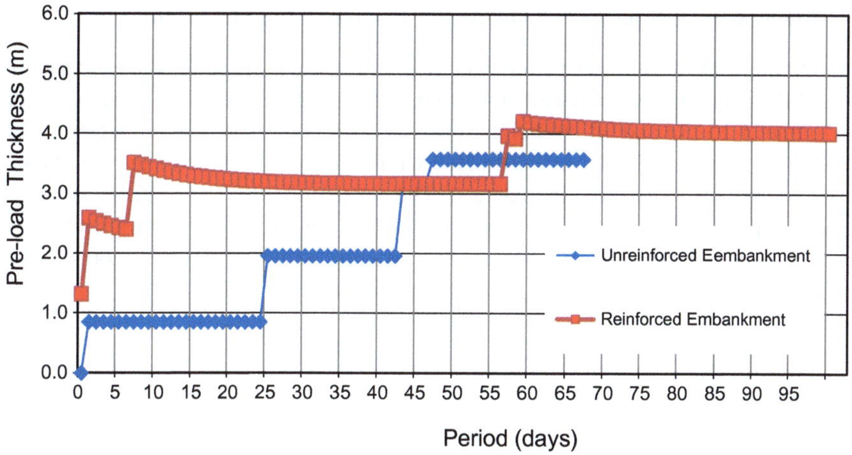

Fig. 3. Stage of construction of the unreinforced and reinforced embankment.

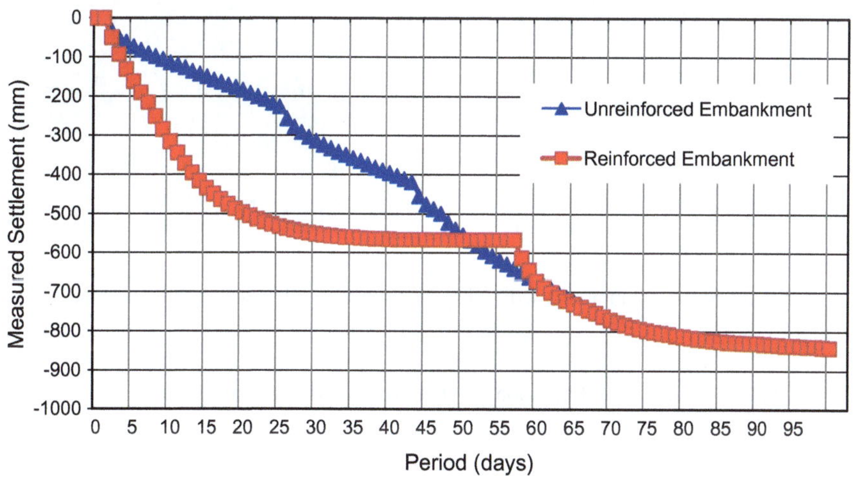

Fig. 4. Settlement measurements of the unreinforced and the reinforced embankments.

3 Numerical Model of Embankment Supported with Bamboo Mattress and Stone Columns

Analyses of the behavior of the reinforced embankment were undertaken by using FEM modeling of PLAXIS. In this model, soft soil of the ground was modeled as undrained soft soil creep (SSC), while the embankment as Mohr-Coulomb (MC) model. The bamboo grid mattress was defined as plate and granular columns as drained MC.

3.1 Mechanical Properties of Geotextile Encased Stone Column

To obtain several mechanical parameters of granular columns such as elastic modulus (E), and Poisson ratio (v), experimental uniaxial tests were undertaken. A stone column with diameter 15 cm and 30 cm height, encased with geotextile, was subjected to axial load (Fig. 5). The stone column was loaded with a load until failure (Table 2). It can be suggested that, based on uniaxial load test. The elastic modulus (E) of the stone column reached at 1317 MPa with Poisson ratio (v) of 0.4. Figure 6 shows the increase axial stress with deformation on the stone column. It must be noted that the geotextile used to reinforce stone column is geogrid with tensile strength of 302 kN/m.

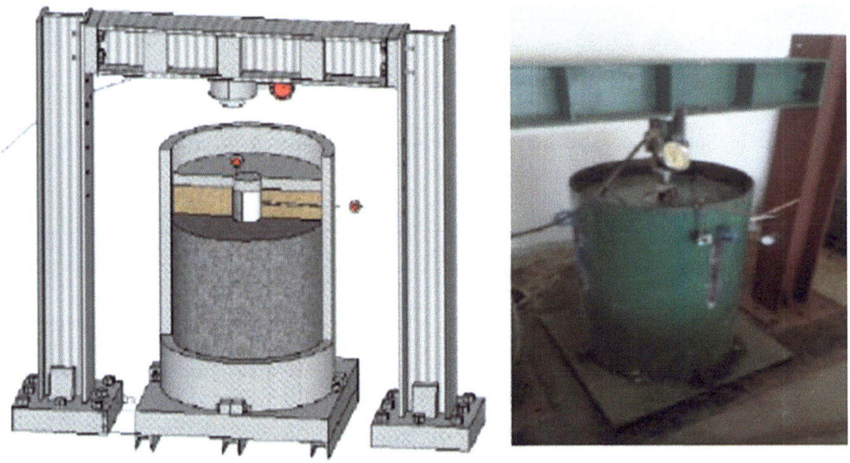

Fig. 5. Stone column encased with geotextile was subjected to uniaxial load.

Table 2. Stress - deformation on stone column during uniaxial load tests.

σ (kN/m^2)	UCS (Mpa)	δ_v (mm)	δ_h (mm)	ε_h
0	0	0	0	0
19,782	1.12	0.0211	0.0030	0.000140
22,961	1.3	0.0261	0.0043	0.000174
22431.38	1.27	0.0267	0.0048	0.000178
21724.88	1.23	0.0267	0.0049	0.000178
21018.38	1.19	0.0266	0.0050	0.000177

3.2 Mechanical Properties of Bamboo Mattress

Mechanical characteristics of bamboo grid were obtained through loading tests. There were two loading tests with different loading position. First is load point at perpendicular of bamboo rows, and second is a parallel of bamboo rows (Fig. 7). It can be seen in Fig. 8, ultimate load for Bamboo grid ranges from 38 kN to 59 kN, and maximum deformation

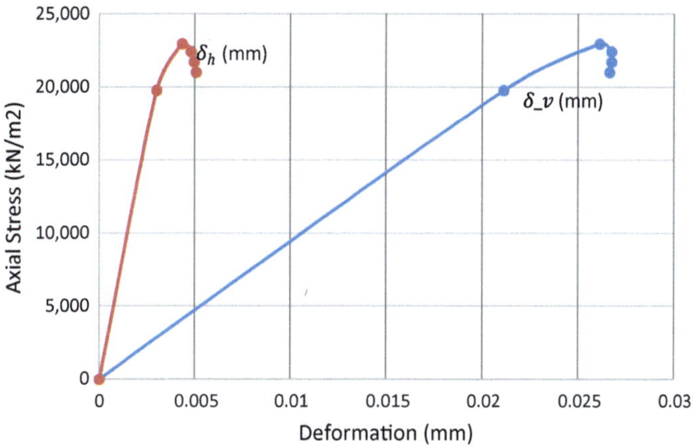

Fig. 6. Axial stress vs deformation on stone column under axial load.

is 97 mm. In terms of elastic modulus of bamboo mattress, it depends on direction of loading. It was found that the perpendicular load on bamboo rows generates larger elastic modulus (715 MPa) than the parallel one (460 MPa). Similar result was found in the sample 2, where 693 MPa is at the perpendicular load pattern compared to 497 MPa at parallel load pattern. Based on the results of bending moment tests, stiffness of bamboo mattress can be computed as seen in Table 3.

Fig. 7. Loading tests on bamboo mattress.

3.3 FEM Model

In the FEM model, the soft soil ground was modelled as silty clay with soft soil creep (SSC) constitutive model, undrained. To compute input parameter in SCC, engineering properties were obtained from Triaxial and Oedometer tests. Based on Coefficient of compressibility (*Cc*) of 1.107 and initial void ratio of 2.987, modified compression index (λ^*) was computed to be 0.12. Swelling index (Cs) of 0.158 can be used for modified swelling index (K^*) at 0.035, and creep index at 0.034 for modified creep

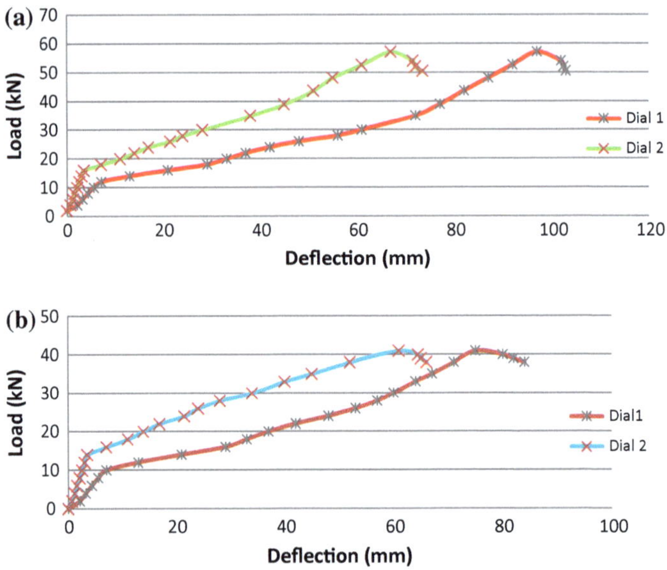

Fig. 8. Load vs deflection at bamboo mattress bending moment test with (a) perpendicular and (b) parallel load patterns.

Table 3. FEM parameter used for bamboo mattress.

Material	Type model	Diameter (m)	EI (kN m^2/m)	EA (kN/m)	W (kN/m/m)
Bamboo mattress	Plate	0.4	5563.75	414562.56	2.640

index (μ*) at 0.003. The parameter used for FEM for other materials including stone column, embankment, and bamboo mattress can be seen Table 4.

Since stone column can behave as reinforcement as well as vertical drain, the stone column can be modelled as stone column with vertical drain inside the stone column. To model stone columns as drain, equivalent horizontal drain was then computed by using Hird et al. (1992) equation which is used to transform the insitu 3D unit-cell axisymmetric condition into equivalent 2D multidrain plane strain condition. As a result, horizontal soil permeability at 1×10^{-4} m/day was converted to 9.98×10^{-5} m/day, assuming no smear zone around the stone column. This value is equivalent to the 6.65×10^{-5} m/day horizontal permeability by considering smear zone (Hird et al. 1992). It is noted that the equivalent horizontal permeability without smear zone is larger than that with smear zone. This indicated that the convertion of horizontal permeability in this study is still reasonable.

To validate the FEM model, the similar stage of construction in the field test of embankment without reinforcement was simulated. It was found that the FEM embankment model generates closed result of the simulated settlement (70.4 cm) to the measured settlement (74 cm) in the field (Fig. 9). For comparison, silty clay model with Hardening Soil (HS) was also modelled with several parameters such as E_{50}^{ref} of

Table 4. Input parameter for FEM of embankment, and stone column, and bamboo mattress.

Stone column		
Constitutive models	MC	[-]
Drainage	drained	[-]
γ_{unsat}	17.94	[kN/m^3]
γ_{sat}	20.68	[kN/m^3]
kx	1.0	[m/day]
ky	1.0	[m/day]
ϕ	50	[°]
c	1.0	[kN/m^2]
E	1317	[kN/m^2]

Embankment		
Constitutive models	MC	[-]
Drainage	drained	[-]
γ_{unsat}	17.00	[kN/m^3]
γ_{sat}	18.00	[kN/m^3]
kx	1.0	[m/day]
ky	1.0	[m/day]
ϕ	30	[°]
c	1.0	[kN/m^2]
E	8000	[kN/m^2]

Bamboo Grid		
Model	Plate	
EI	5563.75	[kNm^2/m]
EA	414562.5	[kN/m]
t	0.4	[m]
w	2.64	[kN/m/m]

4300 kPa, E_{oed}^{ref} of 4300 kPa, E_{ur}^{ref} of 14,400 kPa, m of 0.9, and K_0^{nc} of 0.546. The simulations yield the simulated settlement at 70 cm, closed to the measured settlement at 74 cm (Table 5). Overall, the FEM models generate well agreement results to the field test model, indicating that the FEM models is are reliable, particularly for the FEM model with silty clay SSC.

The Effect of Stone Column and Bamboo Mattress on Settlement of Embankment

The performance of stone column with bamboo mattress in reducing settlement and accelerating consolidation was examined. In this way, the size of the reinforced embankment model is similar to that in the field test. For reinforcement, the embankment was supported with 3 stone columns, in which each column has a 0.6 m diameter and 3 m length. In the interface between the embankment and the top of stone column, a bamboo mattress with a 10 m length was placed (Fig. 9). The parameter of bamboo mattress is shown in Table 3. The filling of 0.9 m thick first embankment was simulated over 7 days, and then followed by a 1.1 m thick second embankment was simulated over 50 days, and 1.2 m thick third embankment was simulated over

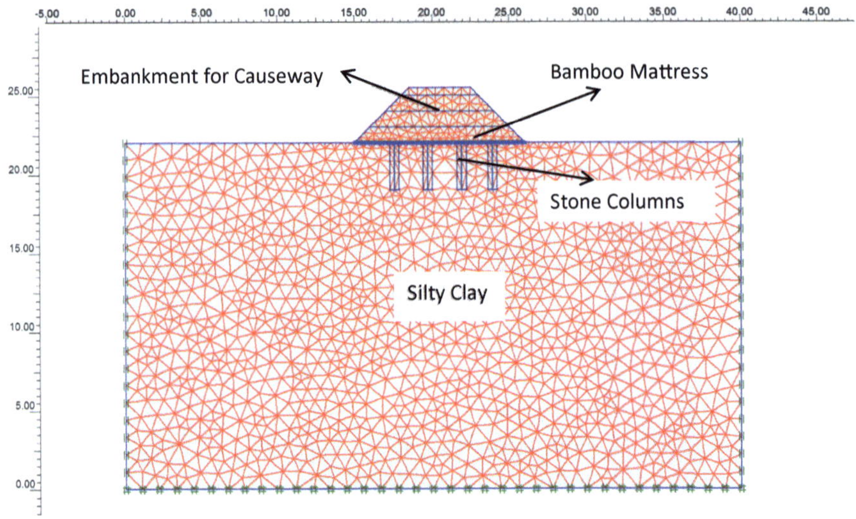

Fig. 9. FEM model of the reinforced embankment with stone columns and bamboo mattress.

Table 5. Comparison of the simulated settlements and the measured settlement for embankments without reinforcement.

Soil	Settlement (m) generated in embankment filling at the period		
	24th day	42nd day	68th day
Silty clay HS	0.02	0.08	0.7
Silty clay SSC	0.03	0.37	0.7
Silty clay in field test	0.226	0.410	0.745

50 days. For comparison, the embankment was simulated to be supported by the stone columns without the interface of bamboo mattress. This aims to examine the effect of bamboo mattress on settlement of the embankment. It can be seen in Table 6, the stone column with bamboo mattress can accelerate settlement 96.4% faster than the unreinforced embankment. In terms of settlement, bamboo mattress and stone columns reinforcement yield settlement slightly lower than the only bamboo mattress reinforcement, 12 cm compared to 14.9 cm. However, their settlements are about a half (54.6%) to the settlement of the unreinforced settlement. This reveals that stone columns contribute mainly to vertical drain of the soil below the embankment while bamboo mattress increase the soil's bearing capacity. The effectiveness of bamboo mattress can be seen in Fig. 10 that the settlement of the embankment reinforced with stone columns minus bamboo mattress yields remain large settlement (18.9 cm) than other methods. This suggested that the bearing capacity generated by the stone columns is not significant increased.

Table 6. Results of the simulated settlements for embankments reinforcement model.

Embankment reinforcement model	Elastic settlement (m)	Consolidation settlement (m)	Total settlement (m)	Period (days)
Unreinforced	0.122	0.104	0.226	2802
Bamboo mattress	0.066	0.083	0.149	2644
Stone columns	0.096	0.093	0.189	977
Stone columns + bamboo mattress	0.018	0.105	0.123	101

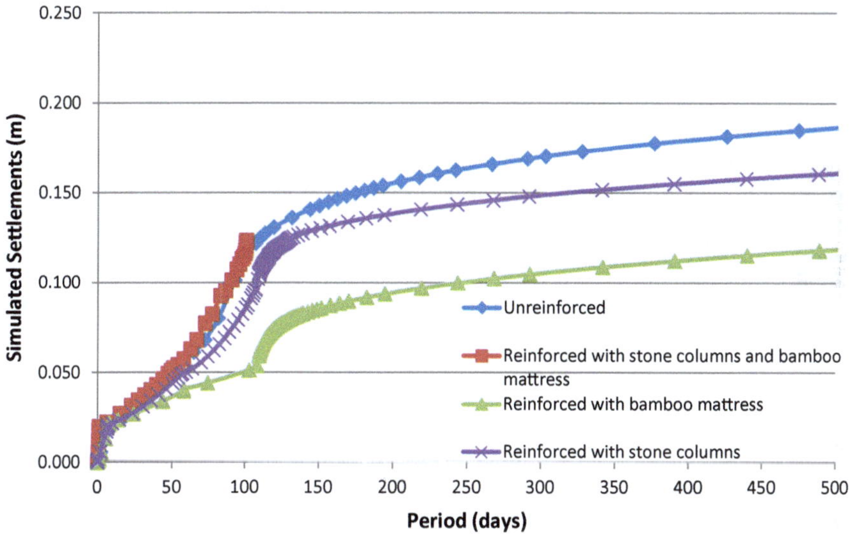

Fig. 10. Simulated settlements generated by the model of embankment with alternative reinforcement method.

4 Conclusions

1. The use of stone columns would accelerate the settlement by 96% while the use of bamboo mattress would increase bearing capacity a double than that of the embankment without reinforcement.
2. The reinforcement of embankment with combination of stone columns and bamboo mattress seems to be more effective compared to the other methods including the reinforcement with bamboo mattress and the reinforcement with stone columns.
3. Stone columns work as vertical drain while bamboo mattress performs to distribute embankment loads over larger area of the soft soil, leading to the increase of the bearing capacity.

Acknowledgements. We would like to express our gratitude to Pelindo IV Makassar Indonesia for providing experiment facility for field tests during the study. We would like to thank for Arwin, Ilham Jafar, Kemal Tharik, and Muhammad Alfa for sharing valuable data and excellent support during experimental test.

References

Abusharar, S.W., Zheng, J.J., Chen, B.B., Yin, J.H.: A simplified method for analysis of a piled embankment reinforced with geosynthetics. Geotext. Geomembr. **27**(1), 39–52 (2009)

Basack, S., Rujiatkamjorn, C.: Analysis of the behaviour of stone column stabilized soft ground supporting transport infrastructure. Procedia Eng. **143**, 347–356 (2016)

Deb, K.: A mathematical model to study the soil arching effects in stone columns-supported embankment resting on soft foundation soil. Appl. Math. Model. **34**(12), 3871–3883 (2010)

Fatahi, B., Basack, S., Premananda, S., Khabbaz, H.: Settlement prediction and back analysis of young modulus and dilation angle of stone columns. Aust. J. Civil Eng. **10**(1), 67–79 (2012)

Guetif, Z., Bouassida, M., Debats, J.M.: Improved soft clay characteristics due to stone column installation. Comput. Geotech. **34**, 104–111 (2007)

Hedge, A., Sitharam, T.G.: Use of bamboo in soft ground engineering and its performance comparison with geosynthetics: experimental studies. J. Mater. Civil Eng. **27**(9) (2015)

Indraratna, B., Basack, S., Rujikiatkamjorn, C.: Numerical solution of stone column improved soft soil considering arching, clogging, and smear effects. J. Geotech. Geoenvironmental Eng. **139**(3), 377–394 (2013)

Low, B.K., Tang, S.K., Choa, V.: Arching in piled embankment. J. Geotech. Eng. **120**(11), 1917–1938 (1994)

A Three Dimensional Discrete Constitutive Model for Over Coarse Grained Soil

Yan Zongling[1,2(✉)], Gou Dongyuan[1,2], and Chai Hejun[1,2]

[1] National Local Joint Engineering Laboratory for Road Engineering and Disaster Prevention and Reduction Technology in Mountainous Areas, Chongqing, People's Republic of China
orienty@163.com
[2] National Engineering & Research Center for Highways in Mountain Area, Chongqing, People's Republic of China

Abstract. Over Coarse Grained Soil is widely used as filling material for subgrade in mountainous highway, and the settlement and stabilization of high embankment settlement is critical to operation safety of mountainous highway. The constitutive relationship of over coarse grained soil is theoretical basis for settlement and stability analysis. So a rigid contact model for over coarse grained soil particles is established combined with the distribution of contact force and normal of contact force based on the research of contact characteristics of particles and ignoring the deformation of particle. The local constitutive model is acquired after analyzing the relationship between contact force of particles and local stress in over coarse grained soil. Furthermore, a three dimensional discrete constitutive model is built up. It is proved through discrete constitutive model that the fabric will change during the deformation of over coarse grained soil which results the change of physical and mechanical characteristics. The change of fabric affects the macro mechanical responses characteristics of over coarse grained soil.

Keywords: Embankment subgrade · Over course grained soils
Discrete granular soil particle · Fabric · Constitutive model

1 Introduction

Over coarse grained soil such as gravel soil, soil-rock aggregate mixture, boulder and rockfill is a typical granular media composed of discrete particles within two or three orders of magnitude in dimension. And it is widely applied in embankment in mountainous highway. The physical and mechanical characteristics is quite different from fine grained soil. Over coarse grained soil is a loose and discrete media in substance, and the contact between particles is discontinuous point-contact. From the paper of Luan and Ugai (1999), for over coarse grained soils and other loose granular media, the research of physical constitutive model of deformation and strength characteristics to overcome inherent limitation of mechanical methods of continuous medium with basis of micro mechanics is quite potential.

S. Wang et al. (eds.), *New Solutions for Challenges in Applications of New Materials and Geotechnical Issues*, Sustainable Civil Infrastructures,
https://doi.org/10.1007/978-3-319-95744-9_8

Fabric is an important concept in mechanics of granular media which can describe dimensional arrangement feature and interaction of granular particles' (Arthur et al. 1977; Mahmood and Mitchell 1974; Oda 1977). Micro fabric of over coarse grained soil is closely related to particle size, shape, rigidity, gradation, void ratio, contact number between particles and stacking manner. Granular media is compose of large number particles, so fabric has statistical significance.

Macro Characteristics of over coarse grained soil can be described by several key micro fabric parameters. Void ratio is a key physical parameter of over coarse grained soil (Oda et al. 1980; Chang 1990). Direction fabric describes the dominant stacking manner of over coarse grained soil particles in three-dimensional space. Particle Pi contacts to neighbor particle P_1 to P_5, and the contact points are C_1 to C_5, contact normal are $n^{(1)}$ to $n^{(5)}$. The relation between Particle Pi and its neighbor particles includes contact point number and contact normal, as shown is Fig. 1.

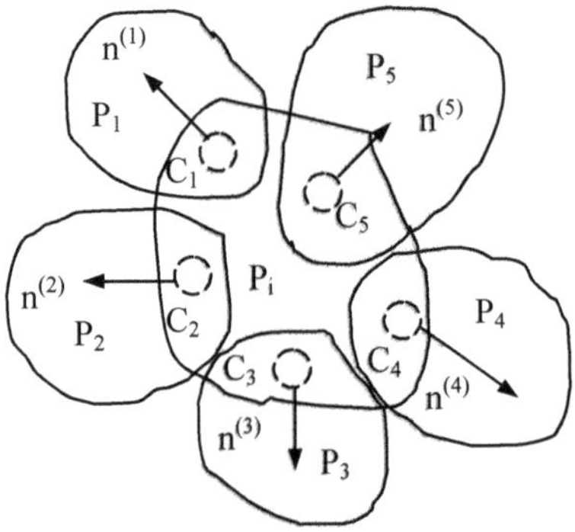

Fig. 1. Particle pi in relation to its neighbor particles

Fabric ellipse is introduced to describe contact normal. Fabric ellipse is a second-order tensor. Probability density function of contact normal $E(\alpha, \beta)$ is closely related to stress ellipse and has specific physical meaning. Probability density function $E(\alpha, \beta)$ changes in loading process so as to change the strength of over coarse grained soil which can result strain-hardening or strain-softening (Oda 1972a, b, c).

It is proved by experiment that probability density function $E(\alpha, \beta)$ can represent the three-dimensional distribution of contact normal (Oda 1972a, b, c). Therefore, particle characteristic of over coarse grained soils, interaction and essential fabric of spatial uniform distribution between particles can be descripted objectively. Based on mathematical function of statistical distribution of the reflected fabric parameters, the relationship could be set up between inherent evolution of microstructure fabric parameters over coarse grained soils and macroscopic mechanics responses.

Granular media mechanics is assumed that over coarse grained soil is composed of solid particles contacted with each other, and interaction between particles obeys the laws of probability.

Granular media mechanics is applied in study the mechanical phenomena on the contact points of particles, and describes the phenomena according to the formula of mathematical statistics. Some mechanical models have been built by different research. However, those models can't interpret the relationship between fabric change and mechanical response of over coarse grained soils under load very well.

2 Rigid Particle Contact Model of Over Coarse Grained Soil

There is a certain gradation within over coarse grained soil particles where large particles act as the skeleton and fine particles fill in the void between the large particles. Both the interaction between particles and between the filled surrounding medium should be belong to solid contact mechanics in terms of the mechanical characteristics (Stake 1983). It is supposed that the number of over coarse grained soil particles is enormous amount, so the macro mechanical parameters have statistical significance. The uneven stress of microcosmic can be described by average stress (Chang 1988a, b). When the over coarse grained soil particle number is infinity, and over coarse grained soil is continuous in the certain space, the summation in the calculation could be changed into integral (Zhong and Yuan 1992).

The particle contact normal density distribution function $E(\vec{n})$ is introduced, where $\vec{n}$ is the contact unit normal vector. The number of contact points within the $\vec{n} \rightarrow \vec{n}+d\vec{n}$ is $E(\vec{n})d\vec{n}$. Supposing $f_i(\vec{x}^{\alpha},\vec{n})$ is the contact force component located at the contact point $\vec{x}^{\alpha}$ with normal $\vec{n}$. For a certain volume V of the over coarse grained soil, the total contact vectors is zero based on the conditions of static equilibrium.

$$\int f_i(\vec{n})E(\vec{n})d\vec{n} = 0 \quad (i = x, y) \tag{1}$$

where $f_i(\vec{n})$ is the average of contact force within $\vec{n}+d\vec{n}$.

The contact between particles is one of the most fundamental problems in particle mechanics model (Mao 1994). Assumed that the over coarse grained soil particles are rigid, and the deformation of particles is ignored, so the contact of over coarse grained soil particles is elastic contact. Over coarse grained soil appears sliding or rolling possibly under load, so that two adjacent particles can be viewed as two rigid body connected by deformed springs on the contact point (Mindlin et al. 1953). Therefore, the deformation of particles is transformed into spring deformation under the contact force. Choosing two similar particles of P, Q in the over coarse grained soil, it is shown in Fig. 2.

For particle P, the equilibrium equation can be established.

$$\sum_{\alpha=1}^{m} f_i^{p\alpha}(\vec{x}^{\alpha},\vec{n}')(x_j^{p\alpha} - x_j^{p}) = \sum_{\alpha=1}^{m} f_i^{p\alpha}(\vec{x}^{\alpha},\vec{n}')(x_i^{p\alpha} - x_i^{p}) \tag{2}$$

where M is the contact number of particle p; $\vec{n}'$ is a unit branch vector, and $\vec{n}' = \dfrac{\vec{x}^p - \vec{x}^q}{|L^{pq}|}$, $|L^{pq}|$ is branch length.

Putting the summation of all particles into integral form in over course grained soils.

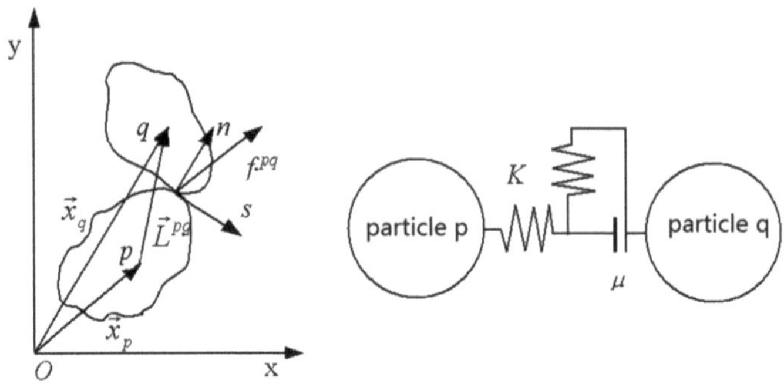

Fig. 2. The rigid contact model of over coarse grained soil

$$\int f_i(\vec{n})L_j(\vec{n}')E(\vec{n})d\Omega = \int f_j(\vec{n})L_i(\vec{n}')E(\vec{n})d\Omega \tag{3}$$

where $L_i(\vec{n}')$ is the average of contact force of the branch vector within $\vec{n} + d\vec{n}$.

Under the contact force, the strain tensor is $\varepsilon_{ij}(\vec{x}^{\alpha})$, the displacement of contact point is $u_i(\vec{x}^{\alpha})$. Ignoring rotation of particles, the equation is shown as below

$$u_i(\vec{x}^{\alpha}) = \varepsilon_{ij}(\vec{x}^{\alpha})L_j(\vec{x}^{\alpha}, \vec{n}') \quad (i, j = x, y) \tag{4}$$

The virtual work of contact force in the unit volume is

$$W = \frac{1}{2V}\sum f_i(\vec{x}^{\alpha}, \vec{n})\varepsilon_{ij}(\vec{x}^{\alpha})L_j(\vec{x}^{\alpha}, \vec{n}') \tag{5}$$

where factor 2 represents each particle counted twice.

The virtual work for average stress is $W = \bar{\sigma}_{ij}\varepsilon_{ij}$.

From Eqs. (2), (5), considering the symmetry of the stress tensor, the equation is obtained as follow.

$$\bar{\sigma}_{ij} = \frac{1}{4V}\sum_{\alpha=1}^{M} [f_i(\vec{x}^{\alpha}, \vec{n})L_j(\vec{x}^{\alpha}, \vec{n}') + f_j(\vec{x}^{\alpha}, \vec{n})L_i(\vec{x}^{\alpha}, \vec{n}')] \tag{6}$$

The integral form is

$$\bar{\sigma}_{ij} = \frac{M}{4V}\int [f_i(\vec{n})L_j(\vec{n}) + f_j(\vec{n})L_i(\vec{n})]E(\vec{n})d\vec{n} \tag{7}$$

Equation (7) links the macroscopic average stress tensor and microscopic first-order tensor together. The over coarse grained soil particle size can be classified into 2 classes, and each class has an average particle size $\bar{d}_k$ with N_k particles. Assuming that all the over coarse grained soil particle size is $\bar{d}$, then

$$V_s = \sum_{k=1}^{l} N_k \cdot \frac{\pi \bar{d}_k^3}{6} = \sum_{k=1}^{l} \frac{\pi \bar{d}_k^3}{6} \cdot \frac{M_k}{\bar{m}}$$
$$= \frac{M}{\bar{m}} \sum_{k=1}^{l} \frac{\pi \bar{d}_k^3}{6}, \quad \bar{d} = \left(\sum_{k=1}^{l} \bar{d}_k^3 \right)^{1/3},$$

So

$$\bar{\sigma}_{ij} = \frac{3\bar{m}}{2(1+e)\pi} \left(\sum_{k=1}^{l} \bar{d}_k^2 \right)^{-\frac{2}{3}} \int_0^{\pi} \int_0^{2\pi} [f_i(\vec{n}) n_j + f_j(\vec{n}) n_i] E(\vec{n}) \mathrm{Sin}\gamma \mathrm{d}\gamma \mathrm{d}\beta \tag{8}$$

where $\bar{m}$ is the average coordination number; e is void ratio, i, j = 1, 2, 3.

For three dimensional form, the particle contact force density distribution function could be described as follow:

$$f_i E(\vec{n}) = C_i + C_{ij} n_j + C_{ijk} n_j n_k + C_{ijkl} n_j n_k n_l + \cdots \tag{9}$$

Omitting the higher order term, then

$$f_i E(\vec{n}) = C_i + C_{ij} n_j \tag{10}$$

And

$$\int_{\Omega} (C_i + C_{ij} n_j) E(\vec{n}) dn = 0 \tag{11}$$

So

$$C_i = 0 \tag{12}$$

Therefore

$$f_i E(\vec{n}) = C_{ij} n_j \tag{13}$$

By substitution formula (13) into the formula (8)

$$\bar{\sigma}_{ij} = \frac{4\bar{m}}{(1+e)} \left(\sum_{k=1}^{l} \bar{d}_k^3 \right)^{-\frac{2}{3}} C_{ij} \tag{14}$$

The three dimensional relationship between integral contact force and stress is

$$f_i = \frac{(1+e)\left(\sum_{k=1}^{l} \bar{d}_k^3\right)^{\frac{2}{3}}}{4\bar{m}E(\vec{n})}\bar{\sigma}_{ij}n_j \tag{15}$$

$E(\vec{n})$ can be expressed as

$$E(\vec{n}) = \frac{1}{4\pi}N_{ij}n_i n_j \tag{16}$$

Substituting formula (16) into formula (15), and the relationship between integral contact force and stress is

$$f_i = \frac{\pi(1+e)\left(\sum_{k=1}^{l} \bar{d}_k^3\right)^{\frac{2}{3}}}{2\bar{m}N_{ij}n_i n_j}\bar{\sigma}_{ij}n_j \tag{17}$$

Assuming that the contact force is f_i (i = n, s, t; n, s, t as the local coordinate), then the relation between force and displacement on the contact point can be expressed as in incremental form.

$$\Delta f_i = D_{ij}\Delta U_j \tag{18}$$

where D_{ij} is contact stiffness tensor.

In local coordinates n, s, t, D_{ij} can be expressed as

$$D_{ij} = D_n n_i n_j + D_s s_i s_j + D_t t_i t_j \tag{19}$$

For macroscopic isotropic over coarse grained soil, the tangential contact stiffness is isotropic on the particle contact plane, so

$$D_{ss} = D_{tt} = D_s \tag{20}$$

Generally, the normal contact stiffness is a function of the normal contact force.

$$D_n = C_1 f_n^{\beta} \tag{21}$$

where C, β functions are associated with the over coarse grained soil characteristic, particle size and surface roughness.

The tangential stiffness is

$$D_s = C_2 D_n \left(1 - \frac{f_s}{f_n \tan\varphi_\mu}\right)^{\eta} \tag{22}$$

where C_2, η coefficients are related to the material itself; ϕ_μ is frictional angle between particles.

3 Three Dimensional Discrete Constitutive Model

Over coarse grained soil particles are randomly stacked, so it can be described by fabric density distribution function. According to findings of Rothenburb and Bathurst (1988) and Bathurs and Rothenburb (1988), the density distribution function of over coarse grained soil particle fabric content could be described approximately by three function in the two dimensional case.

Contact normal vector

$$E(\theta) = \frac{1}{2\pi}(1 + a_1(\cos\theta - \theta_1))] \tag{23}$$

Normal contact force

$$f_n(\theta) = f_0[1 + a_2 \cos 2(\theta - \theta_2)] \tag{24}$$

Tangential contact force

$$f_t(\theta) = -f_0 a_3 \sin 2(\theta - \theta_3) \tag{25}$$

where the minus is negative tangential contact force to rotate counter clock wise positive; $\alpha_1, \alpha_2, \alpha_3$ are coefficients reflecting the degree of anisotropy; $\theta_1, \theta_2, \theta_3$ denote fabric shaft angle, maximum normal contact force, average contact angle of maximum tangential contact force respectively.

f_0 is average tangential contact force.

$$f_0 = \int_0^{2\pi} f_n(\theta) d\theta \tag{26}$$

Substituting formula (23–25) into the formula (8)

$$\begin{aligned} \bar{\sigma}_{11} = \frac{\bar{m} f_0}{(1+e)\pi} \left(\sum_{k=1}^{l} \bar{d}_k^2 \right)^{-1/2} & [1 + \frac{1}{2}(a_1 \cos 2\theta_1 + a_2 \cos 2\theta_2 + a_3 \cos 2\theta_3) \\ & + \frac{a_1 a_2}{2}(\cos 2\theta_1 \cos 2\theta_2 + \sin 2\theta_1 \sin 2\theta_2)] \end{aligned} \tag{27}$$

$$\begin{aligned} \bar{\sigma}_{22} = \frac{\bar{m} f_0}{(1+e)\pi} \left(\sum_{k=1}^{l} \bar{d}_k^2 \right)^{-1/2} & [1 - \frac{1}{2}(a_1 \cos 2\theta_1 + a_2 \cos 2\theta_2 + a_3 \cos 2\theta_3) \\ & + \frac{a_1 a_2}{2}(\cos 2\theta_1 \cos 2\theta_2 + \sin 2\theta_1 \sin 2\theta_2)] \end{aligned} \tag{28}$$

$$\sigma_{12} = \frac{\bar{m} f_0}{(1+e)\pi} \left(\sum_{k=1}^{l} \bar{d}_k^2 \right)^{-1/2} (a_1 \sin 2\theta_1 + a_2 \sin 2\theta_2 + a_3 \sin 2\theta_3) \tag{29}$$

Stress is associated with void ratio, the number of contact points and fabric contents. If the relationship between contact force and contact displacement is linear, the formula is obtained as follow by Hooke's law.

$$\Delta\sigma = E\Delta\varepsilon \tag{30}$$

The tangential stiffness

$$D_s = \lambda D_n \tag{31}$$

where λ is a constant, generally $\lambda = 0.1 - 1.0$.

The relationship between strain and stress of over coarse grained soil can't be derived directly, and the strain should be linked to contact force. Over coarse grained soil relationships between stress and strain can be derived when the local constitutive relation is confirmed.

The contact stiffness tensor is

$$D_{ij} = D_n n_i n_j + D_s s_i s_j \quad (i,j = 1,2) \tag{32}$$

Local contact constitutive relation of incremental form

$$\Delta f_i = D_{ij}\Delta u_j \tag{33}$$

where Δf_i is incremental contact force on the contact point; Δu_j is incremental displacement on the contact point.

For the over coarse grained soil, the number of particles is homogeneous on the large scale, and its displacement is linear distribution without forming shear zone.

$$\Delta u_j = l_i \Delta\varepsilon_{ij} = l(\vec{x}^{\alpha}) n_i \Delta\varepsilon_{ij} \tag{34}$$

where $\Delta\varepsilon_{ij}$ is strain increment.

The relation between contact stress increment and strain increment is

$$\Delta f_i(\vec{x}^{\alpha}, \vec{n}) = l(\vec{x}^{\alpha})(D_n n_i n_j + D_s s_i s_j) n_k \Delta\varepsilon_{kj}, (i,j = 1,2) \tag{35}$$

Substituting formula (35) into formula (34), the incremental formula of force and strain is

$$\Delta\sigma_{ij} = A_{ijkl}\Delta\varepsilon_{ij}, \quad (i,j = 1,2) \tag{36}$$

where

$$A_{ijkl} = \frac{2\bar{m}}{(1+e)\pi}\int_0^{2\pi} (n_i n_j n_k n_l + B_{ijkl} D_s) E(\theta) d\theta \tag{37}$$

$$B_{ijkl} = (n_i s_j n_k s_l + n_j s_i n_k s_l + n_i s_j n_l s_k + n_j s_i n_l s_k)/4 \tag{38}$$

$$E_{ijkl} = \frac{1}{4}(n_i t_j n_k t_l + n_j t_i n_k t_l + n_i t_j n_l t_k + n_j t_i n_l t_k) \tag{39}$$

It is clear that stiffness tensor conforms to symmetry of stress tensor and strain tensor, then

$$A_{ijkl} = A_{jikl} = A_{klij} \tag{40}$$

For three dimensional condition, density function is

$$E(\vec{n}) = \frac{1}{4\pi} \tag{41}$$

Substituting formula (41) into formula (40), and integral is

$$\begin{bmatrix} \Delta\sigma_{xx} \\ \Delta\sigma_{yy} \\ \Delta\sigma_{zz} \\ \Delta\sigma_{xy} \\ \Delta\sigma_{xz} \\ \Delta\sigma_{yx} \end{bmatrix} = \begin{bmatrix} D_{11} & & & & & \\ D_{21} & D_{22} & & & & \\ D_{31} & D_{32} & D_{33} & & & \\ 0 & 0 & 0 & D_{44} & & \\ 0 & 0 & 0 & 0 & D_{55} & \\ 0 & 0 & 0 & 0 & 0 & D_{66} \end{bmatrix} \begin{bmatrix} \Delta\varepsilon_{xx} \\ \Delta\varepsilon_{yy} \\ \Delta\varepsilon_{zz} \\ \Delta\gamma_{xy} \\ \Delta\gamma_{xz} \\ \Delta\gamma_{yz} \end{bmatrix} \tag{42}$$

where

$$\begin{aligned} D_{11} &= D_{22} = \frac{D}{5}(12D_n + 3D_s + 5D_t) \\ D_{33} &= \frac{D}{5}(12D_n + 8D_s) \\ D_{44} &= \frac{D}{5}(4D_n + D_s + 5D_t) \\ D_{55} &= D_{66} = \frac{D}{10}(8D_n + 7D_s + 5D_t) \\ D_{12} &= D_{21} = \frac{D}{5}(4D_n + 3D_s - 5D_t) \\ D_{31} &= \frac{4D}{5}(D_n - D_s) \\ D_{32} &= D_{23} = \frac{4D}{5}(D_n - D_s) \\ D &= \frac{\bar{m}}{8(1+e)\pi\bar{r}} \end{aligned}$$

For isotropous fabric, there is

$$N_{ij} = \begin{cases} 1 & i = j \\ 0 & i \neq j \end{cases} \tag{43}$$

And the density function obeys the formula

$$\int_{\Omega} E(\vec{n})d\Omega = 1 \tag{44}$$

So

$$N_{11} + N_{22} + N_{33} = 3 \tag{45}$$

Stiffness matrix can be expressed as

$$\begin{bmatrix} N_{11} & N_{12} & N_{13} \\ N_{21} & N_{22} & N_{23} \\ N_{31} & N_{32} & N_{33} \end{bmatrix} = \begin{bmatrix} 1 & 0 & 0 \\ 0 & 1 & 0 \\ 0 & 0 & 1 \end{bmatrix} \rightarrow$$

$$\begin{bmatrix} C_{11} & & & & \text{symmet -} & \\ C_{21} & C_{22} & & & \text{rical} & \\ C_{31} & C_{32} & C_{33} & & & \\ C_{41} & C_{42} & C_{43} & C_{44} & & \\ C_{51} & C_{52} & C_{53} & C_{54} & C_{55} & \\ C_{61} & C_{62} & C_{63} & C_{64} & C_{65} & C_{66} \end{bmatrix} = \begin{bmatrix} Q & & & & & \\ S & Q & & & \text{symmet -} & \\ S & S & Q & & \text{rical} & \\ 0 & 0 & 0 & R & & \\ 0 & 0 & 0 & 0 & R & \\ 0 & 0 & 0 & 0 & 0 & R \end{bmatrix} \tag{46}$$

There are three equivalent coefficient for stiffness tensor.

$$Q = \frac{\bar{m}}{10(1+e)\pi\bar{r}}(3D_n + 2D_s) \tag{47}$$

$$S = \frac{\bar{m}}{10(1+e)\pi\bar{r}}(D_n - D_s) \tag{48}$$

$$R = \frac{Q - S}{2} = \frac{\bar{m}}{20(1+e)\pi\bar{r}}(2D_n + 3D_s) \tag{49}$$

There are only two independent coefficient in 12 coefficient, it indicates that the stress-strain characteristics of over coarse grained soil is isotropous when its fabric is isotropous.

The average bulk modulus $\bar{K}$, shear modulus $\bar{G}$ and Young modulus $\bar{E}$ are respectively

$$\bar{K} = S + \frac{2}{3}R = \frac{\bar{m}}{6(1+e)\pi\bar{r}}D_n \tag{50}$$

$$\bar{G} = \frac{\bar{m}}{20(1+e)\pi\bar{r}}(2 + 3\xi) \tag{51}$$

$$\bar{E} = \frac{\bar{m}}{2(1+e)\pi\bar{r}}\frac{2+3\xi}{4+\xi} \tag{52}$$

Poisson's ratio

$$\upsilon = \frac{1-\xi}{4+\xi} \tag{53}$$

It is proved that the bulk modulus and shear modulus of over coarse grained soil is related to void ratio and coordination number.

The bulk modulus is only related to the normal contact stiffness, shear modulus is both related to the normal and tangential contact stiffness, Poisson's ratio is only related to contact stiffness, coordination number, void ratio and particle size, but other parameters are related to particle size and its' distribution, the relationship are shown as Figs. 3, 4, 5 and 6.

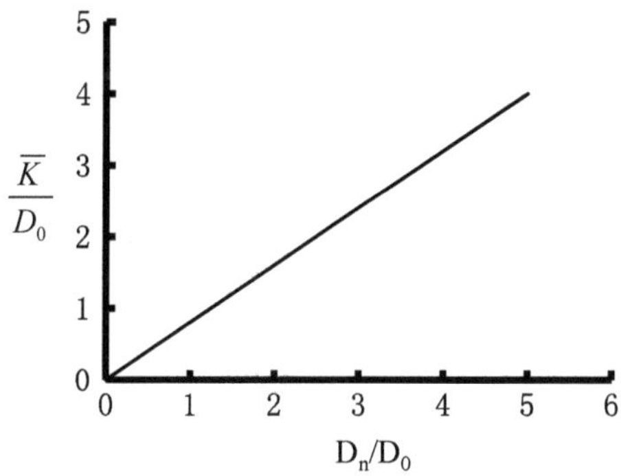

Fig. 3. The relationship between bulk modulus and normal contact stiffness ratio

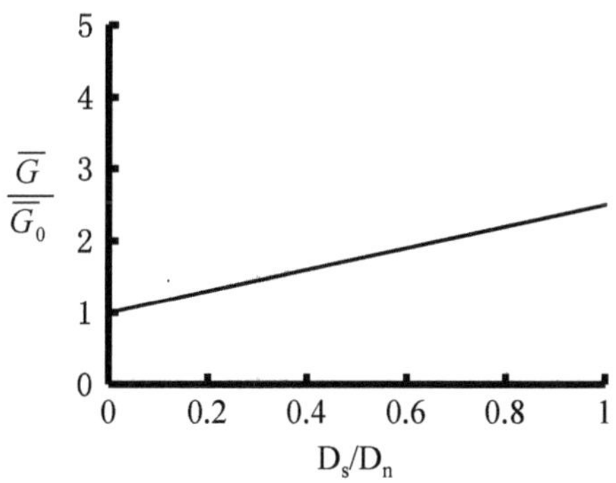

Fig. 4. The relationship between shear modulus and contact stiffness ratio

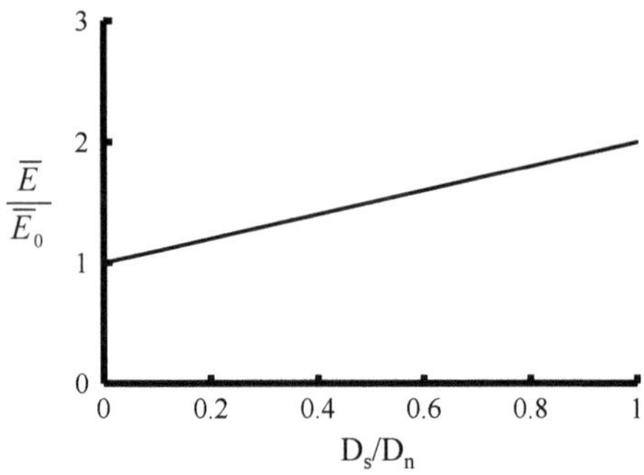

Fig. 5. The relationship between Young modulus and contact stiffness ratio

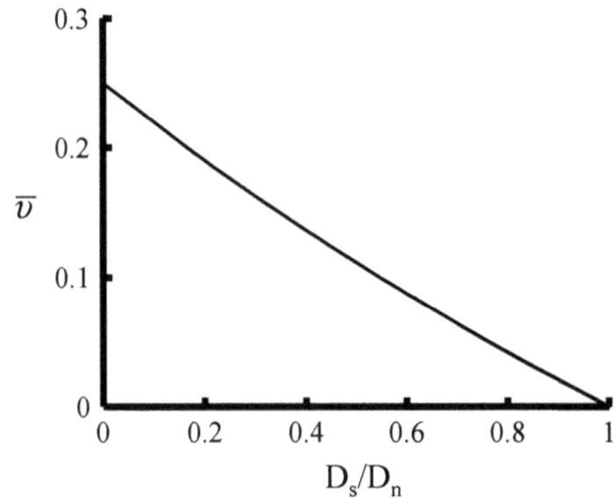

Fig. 6. The relationship between Poisson's ratio and contact stiffness ratio

4 Conclusions

The physical and mechanical properties of over coarse grained soil are closely related to particle spatial stacking mode, void ratio and its' spatial distribution. The relationship between average force and contact point is set up by analyzing the force between particles and normal direction of contact based on the fabric characteristics of over coarse grained soil. Due to fabric change during the process of deformation, relationship between force and strain is nonlinear as same as relationship between stress and contact force, the relationship between stress and contact force is affected by the fabric. The mechanical response of over coarse grained soil in the large scale proved that the changes of fabric contents play an important role on the deformation characteristics of materials.

The rigid contact model of over coarse grained soil is established based on the microstructure of over coarse grained soil particles. Local constitutive relation of over coarse grained soil is acquired by the connection between contact forces of local stress between particles. Based on the local constitutive relation of over coarse grained soil, the constitutive relation of three dimensional granular of over coarse grained soil has been set up. These studies provide foundations for the further research.

Particle flow code (PFC 3D) method is a feasible method for experimental verification of three Dimensional Discrete Constitutive Model. The author will write other paper to discuss.

The application of basic theories of discrete mechanics and particle mechanics is a promising method to study stress-strain characteristics of over coarse grained soil. And the contact force of over coarse grained soil particles and contiguous normal distribution need further research.

Acknowledgements. The authors gratefully acknowledge the financial supports from the National Science and Technology Support Plan under Grant No. 2015BAK09B01, from the National key research and development plan under Grant No. 2016YFC0802203, from the Natural Science Foundation Project of Chongqing Science & Technology Commission of China under Grant No. CSTC2013jcyj A300081, respectively.

References

Luan, M.T., Ugai, K.: Thinking on some fundamental mechanics issues in geotechnical engineering. J. Dalian Univ. Technol. **39**(2), 309–316 (1999)

Arthur, J.R.F., Dunstan, T., Al-Ani, Q.A.J.L., Assadi, A.: Plastic deformation and failure in granular media. Géotechnique **27**(1), 53–74 (1977)

Mahmood, A., Mitchell, J.K.: Fabric-property relationships in fine granular materials. Clays Clay Miner. **22**(5), 197–208 (1974)

Oda, M.: Co-ordination number and its relation to shear strength of granular materials. Soil Fdn. **12**(2), 29–42 (1977)

Oda, M., Konishi, J., Nemat-Nasser, S.: Some experimentally based fundamental results on the mechanical behaviour of granular materials. Géotechnique **30**(4), 479–495 (1980)

Chang, C.S., Ma, L.: Modeling of discrete granulates as micropolar continua. J. Eng. Mech. **116** (12), 2703–2721 (1990)

Oda, M.: The mechanism of fabric changes during compression deformation of sand. Soil Fdn. **12**(2), 1–18 (1972)

Oda, M.: Initial fabrics and their relations to mechanical properties of granular materials. Soil Fdn. **12**(1), 17–36 (1972a)

Oda, M.: Deformation mechanism of sand in triaxial compression test. Soil Fdn. **12**(4), 45–63 (1972b)

Stake, M.: Fundamental quantities in the graph approach to granular materials. In: Jechins, J.T., Stake, M. (eds.) Mechanics of Granular Materials, pp. 9–20 (1983)

Chang, C.S.: Strain tensor and deformation for granular material. J. Eng. Mech. **116**(4), 790–804 (April 1988)

Chang, C.S.: Micromechanical modelling of constitutive relation for granular materials. In: Stake, Jenkin J.T. (eds.) Micromechanics of Granular Materials, pp. 271–278 (1988)

Mao, J.: Preliminary study of mechanical model and calculation method of stress for granular medium. In: Proceedings of geotechnical engineering mechanics and its application symposium for young scholars of China. Wuhan (1994)

Zhong, X., Yuan, J.: Microfabrics and constitutive relations of granular materials. Chinese J. Geot. Eng. **14**(S1), 39–48 (1992)

Mindlin, R.D., et al.: Elastic sphere in contact under varying oblique forces. J. Appl. Mech. Eng. **20**(3), 327–344 (1953)
Wang, M., Qian, Q.: A study on the elastoplastic dynamic constitutive law of granular medium. Acta Mech. Solid Sinica **16**(2), 175–180 (1995)
Gladwell, G.M.L.: Contact Problem of Classical Elastic Theory. Peking University Press (1991)
Rothenburb, L., Bathurst, R.J.: Analytical study of induced anisotropy in idealized granular materials. Geotechnique **39**(4), 601–614 (1988)
Bathurs, R.J., Rothenburb, L.: Micromechanical aspects of isotropic granular assemblies with linear contact interaction. J. Appl. Mech. **55**(1), 17–23 (1988)

Optimization Charge Scheme for Multi-row Ring Blasting Design Adopting Equilateral Triangle Layout Based on Modified Harries' Mathematical Model from a Fragmentation Perspective: A Case Study

Mingzheng Wang, Xiuzhi Shi, and Jian Zhou(✉)

School of Resources and Safety Engineering, Central South University, #932 Lushan South Road, Changsha 410083, China
mz_wang@csu.edu.cn, shixiuzhi@263.net, csujzhou@hotmail.com

Abstract. To eliminate undesired phenomena such as brow damage and excessive fines owing to the energy concentration on the collar of the ring, a fragment simulation using a charge scheme optimization algorithm is proposed based on modified Harries model in this article. First, the traversal algorithm for an arrangement of regular holes (holes charged to the collar) is proposed based on the analysis of the current interval charge design method in ring blasting. Second, a new developed mathematical model based on Harries model is introduced to predict the fragment size of the column charge using the superposition method. Then, the optimization criteria based on the calculated blasted fragment matrix are discussed. Finally, a case study is introduced to illustrate the usage and optimization procedure of the program. Ten trial blastings were conducted in the Tonglvshan copper mine to verify the feasibility and effectiveness of the optimization algorithm. The findings reveal that the probability of brow damage was reduced efficiently.

Keywords: Ring blasting · Charge scheme optimization · Fragmentation matrix · Harries' mathematical model

1 Introduction

Fan-pattern holes are widely used in many mining methods owing to economical advances in stoping and high drilling efficiency. Meanwhile, the planned spacing diminishes as it nears the collar; the explosive charge in the sector is always uneven compared to the parallel holes, which may lead to undesired phenomena including hanging roofs, brow damage or over crushing. Moreover, in the design of the underground mine blast, it is common to use double powder factor to ensure fine breakage in the limited space available (Darling 2011). Thus, it is important to prevent adverse phenomena caused by the concentration of energy, especially in the lower part of the ring, by improving the quality of the blasting design and optimizing the charge structure.

S. Wang et al. (eds.), *New Solutions for Challenges in Applications of New Materials and Geotechnical Issues*, Sustainable Civil Infrastructures,
https://doi.org/10.1007/978-3-319-95744-9_9

Many researchers have proposed solutions to these problems. To achieve better use of the explosive energy, Hagen (1987) suggested drilling holes in a fan-pattern based on a reoriented equilateral triangle. Zhang (2011, 2014) suggested decoupling the charge (using cartridge charges in the lower parts of the holes) or moving primers to the toe of holes to reduce tensile stress in the eyebrow region. These modifications can effectively reduce the overcrush effect near the brow due to the concentration of explosive energy. However, a reasonable scheme for the charging and stemming length (as shown in Figs. 1 and 2) is a basic requirement for ring blasting design whenever the mentioned measurements are applied. Therefore, it is necessary to discuss the method used to develop an optimal charge scheme for fan-pattern holes, especially in the region near the brow.

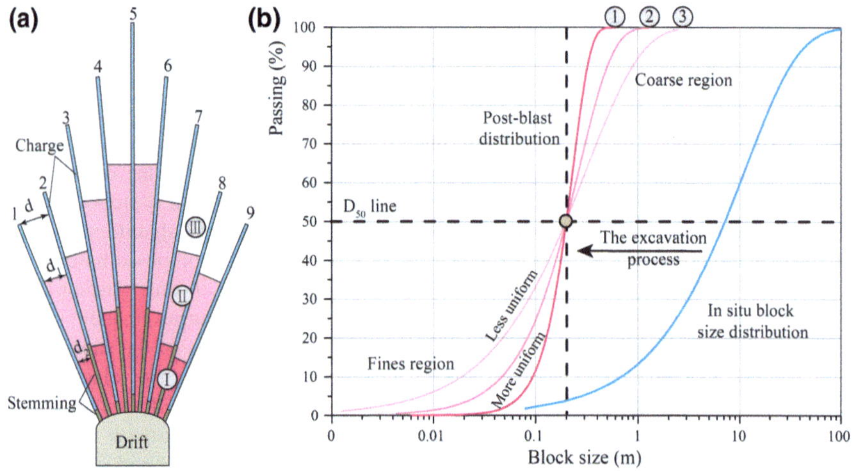

Fig. 1. The influence of the charge scheme on the block size distribution in ring blasting.

The final blasting effect is mainly influenced by two factors. The first is the basic blasting condition, including rock properties, the structure of the rock mass and free face, etc. The second is the design aspect, which covers the drilling pattern, charge structure, explosive properties and detonation method (Scott et al. 2006; Saharan et al. 2006). The design parameters should change in concert with the variation of the initial conditions and should be synchronized with them for a better blasting effect. For mass blasting in underground production, some initial conditions, such as the rock condition and the free face vary notably along the blasting area, which makes it difficult for the engineers to ensure that the design parameters exactly match the given conditions. A mathematical model or simulation tool is needed that can provide a rationale for designers to adjust the design parameters accurately based on the results of numerical simulations.

There are several mature mathematical models of blasting that have been used in the field. For blasting mechanics and process simulation, there are the fracture mechanics blasting model proposed by Margolin and Adams (1982), the KUS damage

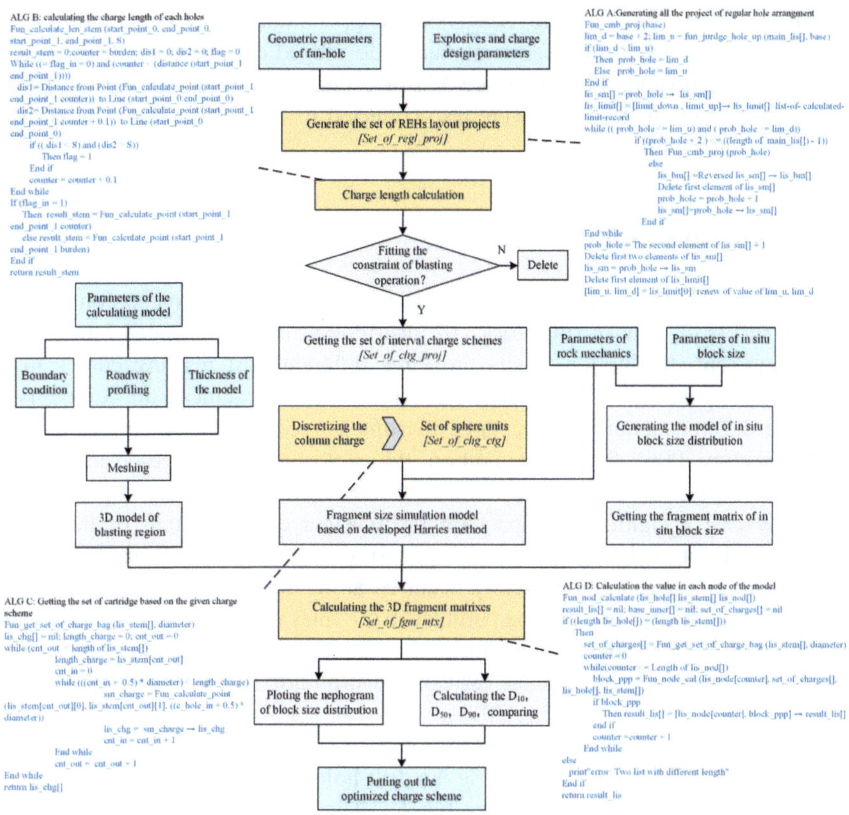

Fig. 2. System flowchart and partial code of the algorithm for the proposed method.

model proposed by Taylor et al. (1986), the PPV damage model (Holmberg 1979; Blair 1999) and the BLAST-CODE model proposed by Qu et al. (2002). For fragment prediction, the Harries model (1983, 1990) can simulate the development of cracks and the fragment size generated by a sphere cartridge. Morin and Ficarazzo (2006) proposed a working form of the Kuz-Ram fragment prediction model based on Monte Carlo simulations. There is a hybrid finite-discrete element model proposed by An et al. (2017) that can simulate the rock crushing, block casting and muck pile profile. However, most mature software products are aimed at the computer-aided design of the layout of the drill-holes, muck pile profile prediction and the prediction of the block size distribution in an open pit mine (Chung et al. 1991; Carter 1992; Liu et al. 2014). There are also many geological data management software packages and cartography systems used in underground mines. However, the functionality of the computer-aided design tools that meet the specific demands of complex mining operations is limited.

This article introduces a new algorithm for calculating the optimal fan-hole charge scheme. First, the current design method for determining the fan-hole charge scheme and its deficiencies are discussed in the context of the rock-crushing mechanism for ring blasting. Second, the methodology of the proposed algorithm is introduced in detail. Then, a case study of the Tonglvshan copper mine is presented and the two different charge scheme projects, one manually developed and the other designed by a program, are compared. Finally, more blasting data were acquired during on-site observation, and production data were also collected to verify the effectiveness of the optimization algorithm.

2 Problem Description

The process of the excavation by blasting described from the fragmentation aspect is illustrated in Fig. 1. The in situ block size distribution is been changed into the post blasting fragment distribution. There are three different post blasting curves in the Figure, where the curve 1 has the most uniform distributed fragment, the curve 3 is the worst. The different quality of the three blasting curves can be influenced by three aspects (Saharan 2006): the rock condition, the explosive and charge design, and the boundary condition. In the underground ring blasting operation, the charge scheme design may cast a significant impact on the blasting effect in light of the limited boundary condition and the layout of fan-pattern drill-hole. As mentioned above, the explosive energy concentration usually occurred at the collar of the ring. Thus, as we can imagine that the resource of fine particles in the fines region in the Fig. 1 is located in the brow of the ring. Besides the fines ore, the brow damage and the back damage are also result from the unreasonable explosive distribution in the centre or the brow of the ring to a much extent (Darling 2011).

The current design method of the ring blasting is kind of randomness and uncertainty (Wang et al. 2017). First, dividing the drill-holes in the ring into several groups, where there are at least three holes in each holes. The holes located on the boundary of each groups are charged to the collar, which means the length of stem is equal to the minimum permitted stem length. Hereinafter refer to as REH (regular hole). Second, the charge length of the other holes in the groups are determined by the maximum spacing principle ensuring that the time delay between holes would not be influenced (Darling 2011). The stem length of these holes are always longer than the minimum permitted stem length. Hereinafter refer to as IRH (irregular hole).

The charge scheme is mainly determined by the arrangement of the REHs. However, it is kind of subjective and random. There are two contrary idea about the arrangement of the REHs:

1. The location of the two adjacent REHs should be arranged as close as possible, ensuring the ample crushing effect by a higher the powder factor.
2. The explosives in the collar of ring should be reduces, thus the spacing of the REHs should be located as far as possible.

Actually, the most of designers' principle is in between these two rules mentioned above, influenced by their experience and the specific condition of different mine.

Besides the interior effect of charge length between different holes in the same ring, the mutual influence of the charge scheme in each rings should also be considered. The reorientated equilateral triangle arrangement of drilling holes between rings can eliminate the adverse effect due to the uneven distribution of holes (Hagen 1987).

However, the charge schemes of two adjacent rings are different. The spatial distribution of explosive must have a significant impact on the holistic blasting effect.

So, these three principles should be considered in charge scheme design for the multi-ring blasting with triangle arranged fan-pattern holes:

1. The charge distribution should keep in step with the variation of the rock condition in different blasting region.
2. The arrangement of REHs in each ring should be optimized.
3. The mutual influence between rings should be considered, the charge scheme design should be optimized according to the geometric parameters of holes in adjacent rings. In underground mining operations, the rock condition, boundary conditions and free surface profile varies considerably with each blasting case, even in the same stope. Using only one type of charge scheme as is current practice is not reasonable. Determining the specific condition of each ring and calculating different charge schemes manually places too much of a workload on the blast designer. Thus, faced with complex and mutable geologic conditions, the issue of how to calculate an optimal charge scheme efficiently will be discussed in this article.

3 Methodology

According to the previous discussion, the issue is concerned with three key points. First, there are many feasible charge schemes that meet the requirements of current basic charge rules for ring blasting. Second, to figure out which one is the most suitable, rational evaluation criteria are needed that can describe each project quantitatively. Third, for the applicability and operability of the design method, the method must not only deal with the complex and mutable geological condition in underground mining but also must not require too much manual work for the designers.

In this article, a computer-aided design and optimization method is proposed; the program is implemented using AutoCAD on a Visual LISP platform. Figure 2 shows the major steps to establish a computer-aided ring blasting design and simulation system. The detailed steps are given as follows.

Step 1. Generate all the possible REH arrangement projects in each rings based on input drill-hole parameters. Then, filter the results using the primary demands of the blasting operation and other customer constraints provided by the designer, and calculate the set of all the feasible REH arrangement projects by traversal algorithm (Wang et al. 2017). Calculate the charge length of IRHs, and acquire the set of all feasible charge schemes. Finally, combining all the projects between two adjacent rings in the minimum calculated unit.

Step 2. Based on the geometric conditions of the blasting area and the accuracy required, generate a 3D meshed model of the blasting region.

Step 3. Calculate the blasted block size value at each node of the 3D model based on the modified Harries mathematical model (Wang et al. 2017) and the in situ block size distribution. Then, acquire the 3D blasted block size value matrices of each feasible charge scheme.

Step 4. Calculate several indexes of each block size matrix to evaluate the quality of each charge scheme. The two key evaluation index of the block size distribution D_{50} and $D_{90} - D_{10}$ are adopted which can be used to measure the centralization and the fluctuation range of the size value. A fragment size nephogram in the 3D blasting model can also be drawn for further analysis to compare the blasting performance of each project.

4 Case Study

4.1 Background and Calculation Model

The Tonglvshan copper mine is located in Daye Hubei Province, producing 800 thousand tons of ore per year, using sublevel stoping as the major mining method. The stopes are laid perpendicular to the trend of the ore body. The length of each stope is equal to the thickness of the vein. The width of each stope is 8.0 m. The height of each level is about 40.0 m. There are three sublevels (with a height of around 13.0 m) in each level. For blasting operations, the diameter of a drill-hole is 64 mm. The space between each ring is 1.6 m in lateral blasting area. The drill-hole toe spacing is 1.9 m, and the minimum permitted stem length at the collar of a drill-hole is 1.2 m.

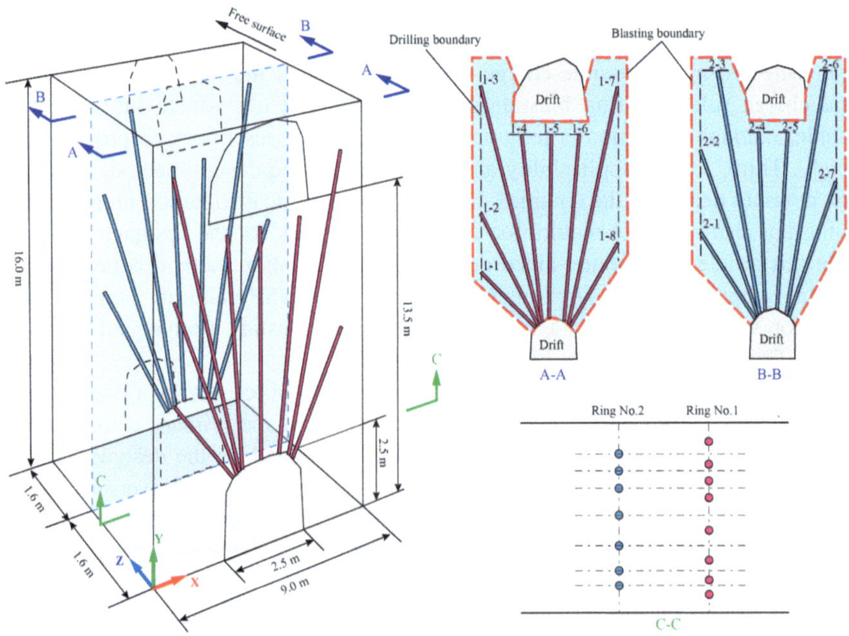

Fig. 3. Geometric properties of the calculated model.

The 9225# room stope at the −450 m level was selected as the trial stope. The geometric properties of the blasting model are illustrated in Fig. 3. The drill-hole parameters, explosive parameters and rock mass properties are given in Tables 1 and 2.

Table 1. Geometric parameters of the fan-holes in the model

Borehole (No.)	Depth (m)	Angle (degree)
1-1	4.5	45
1-2	7.2	63
1-3	13.9	76
1-4	11	84
1-5	10.7	89
1-6	10.9	85
1-7	13.8	78
1-8	5.5	59
2-1	5.9	57
2-2	10	71
2-3	13.5	81
2-4	10.4	88
2-5	10.3	86
2-6	13.3	81
2-7	8.2	70

4.2 Multi-scheme Optimization

Inputting the rock parameters, charge parameters and geometric information of drill-holes, the program put out nine projects for ring No. 1, nine projects for ring No. 2. The result of all the projects are given in Tables 3 and 4. Finally, there are 81 kinds of projects generated in this unit of rings after combination.

Then, the final fragment matrixes of all the combination are calculated. And the two key index of the fragment distribution are calculated. The result is shown in Fig. 4.

Among all the projects whose D_{50} located in the reasonable fluctuate range. The project marked in the figure has the minimum $D_{90} - D_{10}$. Thus, it is determined as the optimal charge project for the two adjacent rings in the unit.

5 Discussion

The optimized project (Hereafter Project A) adopts the charge scheme 2# for ring No. 1, 1# for ring No. 2 according to the program. The specific charge parameters of the charge scheme can be found in Tables 3 and 4. In comparison with the previous blasting design data in the Tonglvshan mine, there is a combination similar with the charge scheme, which the designers were accustomed to use in former blasting operation. Adopting charge scheme 7# for ring No. 1, 9# for ring No. 2. Thus, it is necessary to compare the two charge schemes, Project A and B, the optimal charge scheme and the manually designed scheme.

First, the nephogram of fragment size can be drawn according to the final blasted fragment matrix of each project. The fragment size in the blasting region of Project A and B is shown in Figs. 5 and 6. Figure 5a is the fragment in the whole calculated region, Fig. 5b–e are the fragment nephograms of Project A, starting at the plan

Table 2. Parameters referenced in the paper

Parameters		Value
Rock properties		
v*	Poisson's radio	0.25
ρ*	Density	3.3 g/cm^3
C_P*	Longitudinal wave velocity	5200 m/s
T*	Dynamic ultimate tensile strain of rock	2.0e-4
D_{50situ}*	Mean value of in situ block size distribution	1.0 m
ε	The tangential strain at the wall of hole	
n	Number of cracks	
λ	Block size value	m
λ_0 (i, j, k)	Fragment matrix of in situ block size	
λ_1 (i, j, k)	Fragment matrix of blasting block size	
λ_{fnl} (i, j, k)	Fragment matrix of the final blasted block size	
Explosive parameters		
ρ_1*	Density	1.1 g/cm^3
P*	Pressure of explosion	7500 MPa
K*	Adiabatic index	2.0
Borehole properties and charge parameters		
b*	Radius	0.032 m
d*	Toe spacing of drill-hole	1.9 m
B*	The minimum permitted stem length	1.2 m
S*	Influence radius of column charge	0.95 m
$(D_{50\min}, D_{50\max})$ *	Fluctuation range of reasonable mean size of fragment	(0.15 m, 0.25 m)
D_{10}	Sieve size of 10% passing value	m
D_{50}	Sieve size of 50% passing value (mean value)	m
D_{90}	Sieve size of 90% passing value	m
*	Meshing scale	0.1 m
(μ, θ, h)	Cylindrical coordinates of the calculated point	

Parameters marked with * represent the input parameters in the calculation.

$z = 0$ m, there are five sectors spacing 0.8 m parallel to the ring. Figure 6a is the fragment distribution in the whole calculated model. Figure 6b–e are the nephograms of the optimized charge scheme (Project B). For Project A, the distribution of REHs is more dispersed compared with that of Project B; the explosives are also uniformly distributed in the brow of the ring. This result can be clearly reflected in the fragment size nephogram of the plan $z = 0$ m, seen in Figs. 5b and 6b, and of $z = 0.8$ m, seen in Figs. 5c and 6c. After optimization, fines in the brow of the ring were notably reduced. The optimization has a positive effect on the brow protection and the improvement of the fragment size distribution.

Then, ranking the elements of the final fragment matrix by the size value, the proportion of elements at series size levels raising at a given increment was calculated. Then, the fragment distribution was generated by fitting with the Rosin-Rammler

Table 3. Charge schemes of the ring No. 1 calculated by the program

Charge scheme	Charge length for each holes (m)								
	1-1	1-2	1-3	1-4	1-5	1-6	1-7	1-8	Total
1#	3.3*	3.7	12.7*	5.1	8.3	5.5	12.6*	2.9	54.1
2#	3.3*	4.7	8.0	9.8*	6.4	5.5	12.6*	2.9	53.2
3#	1.9	6.0*	8.0	9.8*	6.4	5.5	12.6*	2.9	53.1
4#	3.3*	4.7	11.5	6.5	9.5*	5.6	12.6*	2.9	56.6
5#	1.9	6.0*	10.4	6.5	9.5*	5.5	12.6*	2.9	55.3
6#	3.3*	3.7	12.7*	5.1	9.5*	5.5	12.6*	2.9	55.3
7#	3.3*	3.7	12.7*	5.1	8.3	9.7*	8.3	4.3*	55.4
8#	3.3*	4.7	8.0	9.8*	6.4	9.7*	8.3	4.3*	54.5
9#	1.9	6.0*	8.0	9.8*	6.4	9.7*	8.3	4.3*	54.4

Holes marked * represent that they are REHs in the specific charge scheme.

Table 4. Charge schemes of the ring No. 2 calculated by the program

Charge scheme	Charge length for each holes (m)							
	2-1	2-2	2-3	2-4	2-5	2-6	2-7	Total
1#	2.6	8.8*	8.9	8.0	6.6	12.1*	3.9	50.9
2#	4.7*	5.4	12.3*	5.3	6.6	12.1*	3.9	50.3
3#	4.7*	6.8	8.3	9.2*	6.6	12.1*	3.9	51.6
4#	2.6	8.8*	8.3	9.2*	6.6	12.1*	3.9	51.5
5#	4.7*	6.8	8.3	9.2*	7.9	9.1	7.0*	53
6#	2.6	8.8*	8.3	9.2*	7.9	9.1	7.0*	52.9
7#	4.7*	6.8	11.1	7.9	9.1*	9.1	7.0*	55.7
8#	2.6	8.8*	9.0	7.9	9.1*	9.1	7.0*	53.5
9#	4.7*	5.4	12.3*	5.3	9.1*	9.1	7.0*	52.9

Holes marked * represent that they are REHs in the specific charge scheme.

model. The fragment size distributions of the two projects are shown in Fig. 7. It can be clearly observed that the fines of Project A were reduced. Although the D_{50} value is increased, the fragment distribution was more concentrated compared with Project B.

Brow damage and undesired fragment quality are frequently seen adverse phenomena in the Tonglvshan copper mine. To justify the validity of the optimization program, further tests were conducted. The results using the computer-designed method and manually designed method were compared, after carrying out 10 blastings using the optimization method. Among the 10 times the computer designed blasting was performed, brow damage appeared 1 times and no bulk appeared. According to the production data in the Tonglvshan mine, the average rate of brow damage or back damage is higher than 30% in previous production, which indicate that the possibility of brow damage can be efficiently reduced due to the more reasonable charge structure. Figure 8a shows the broken brow at the end of the tunnel in 9312# room stope Sublevel No. 3 at the −485 m level; Fig. 8b is a photo of bulk in a blasting case carried out in 9312# room stope Sublevel No. 2 at the −485 m level.

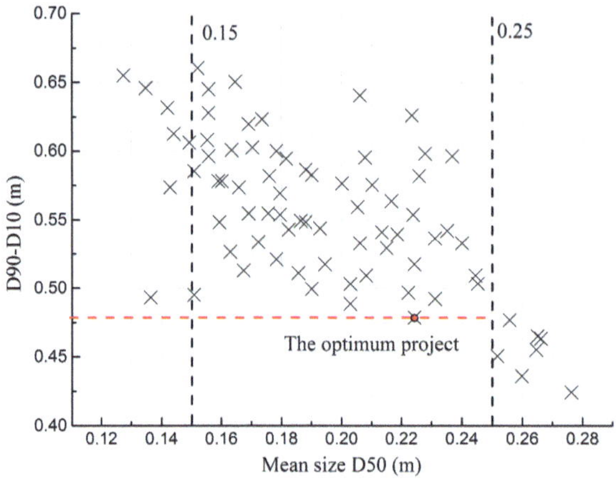

Fig. 4. Results of the evaluation index of the nine feasible charge schemes.

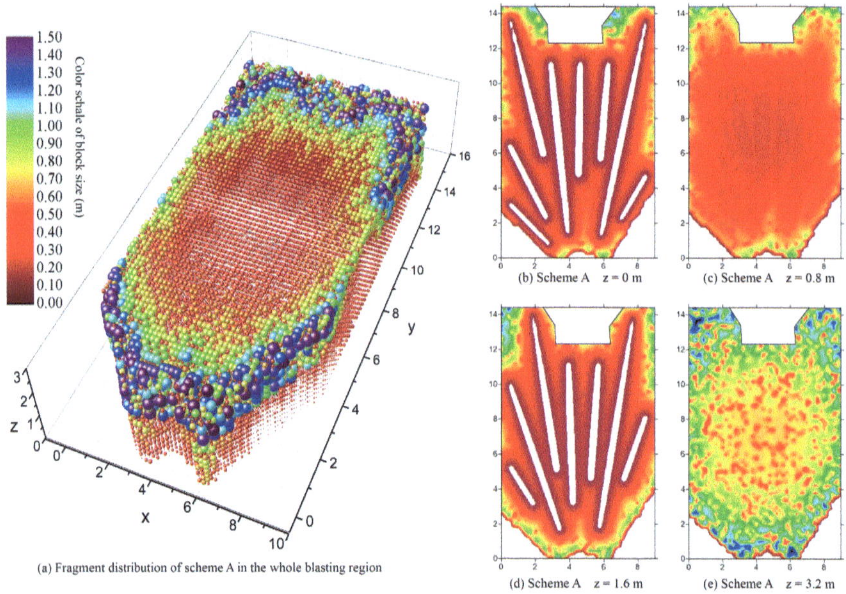

Fig. 5. The nephograms of fragment size in blasting region of the charge schemes A.

The linear superposition method is adopted in the modified Harries model, which considers the linear accumulated effect of sphere units to substitute for the effect of blasting of column charges. However, linear superposition may have some deficiencies (Blair 2008). Thus, to obtain a more rational result regarding the damage effect of column charges based on Harries' model, the algorithm needs to be further improved to find a suitable non-linear superposition function to describe the accumulated damage due to rock blasting. Moreover, delayed initiation is always used in ring blasting to

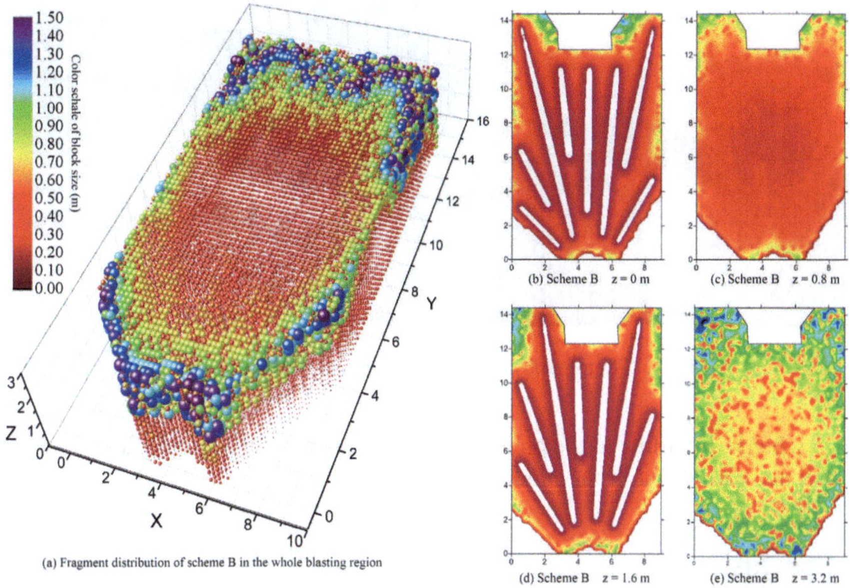

Fig. 6. The nephograms of fragment size in blasting region of the charge schemes B.

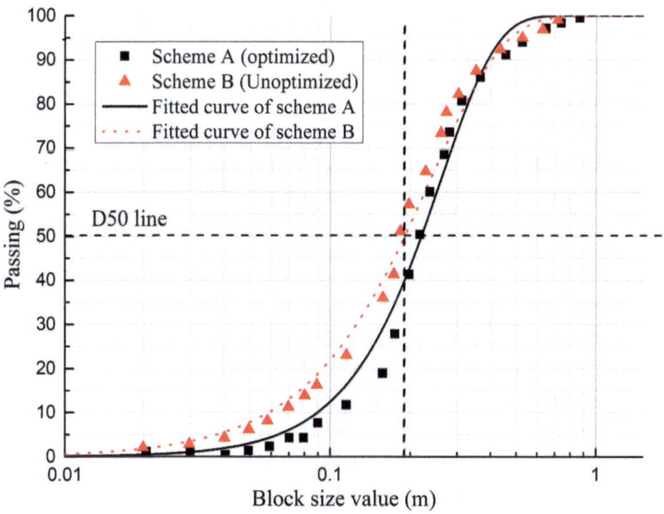

Fig. 7. Comparison of the blasted fragment distribution of the two charge schemes, A and B.

reduce vibration and improve the utilization of energy in the stress wave. However, the delay time is not a major variable controlling fragmentation (Blair 2009). Some small-scale tests (Johansson and Ouchterlony 2013) and numerical simulations (Schill and Sjöberg 2012; Yi et al. 2012) also show that the short delay has no significant impact on fragmentation. The program should achieve a more uniform distributed

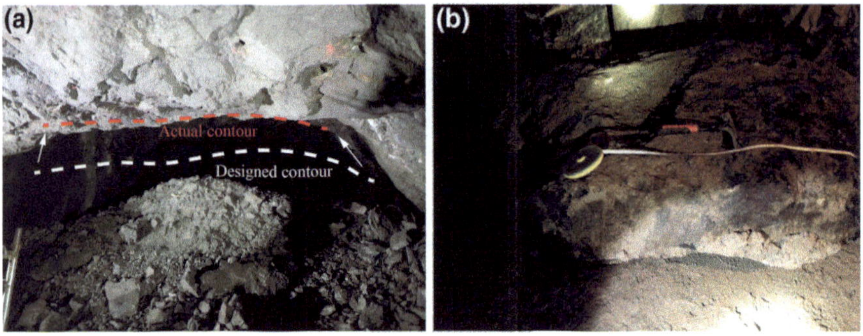

Fig. 8. Photograph of brow damage and bulk in the field.

charge scheme according to the predicted blasted fragment distribution. Thus, the time effect is not considered in the modified Harried model.

6 Conclusion

To achieve the optimum charge scheme at the collar of the ring in ring blasting, a new computer-aided design and fragment simulation optimization program is proposed based on the modified Harries model. By inputting the geometry information of the fan-holes, rock mass properties, explosive parameters and boundary conditions, all the feasible charge schemes and the final blasted fragment matrix of each project can be calculated. The optimum charge scheme can be selected by comparing some key indexes of the fragment matrix and analysing the fragment nephogram of the blasting region. To verify the feasibility of the optimization program, some field blasting tests were conducted in the Tonglvshan copper mine. The following conclusions were reached:

1. By comparing the simulation results of the two charge schemes that were designed manually and via the optimization program, it was found that the optimization program can reduce the explosive concentration on the brow of the ring and that the theoretical fragment distribution can also be improved.
2. In 10 field tests, there is only one time brow damage occurred, where the average rate of brow damage is higher than 30%. This result indicates the optimization program can positively affect brow protection and can improve the quality of the blasting distribution. However, more trial blasting needs to be conducted to verify the reliability and the feasibility of the optimization method.
3. The algorithm proposed in this article is intended to acquire a more uniform distributed charge scheme and to reduce the explosive concentration at the brow ring. The time effect should be further considered about the modified Harries model to obtain a more precise result from the simulation regarding the blasting process and fragment prediction.

Acknowledgements. This research is partially supported by the National Key Research and Development Program (Grant No. 2017YFC0602902) of China, the Shenghua Lieying Program of Central South University (Principle Investigator: Dr. Jian Zhou), and the Fundamental Research Funds for the Central Universities of Central South University (Grant No. 2017zzts567).

References

An, H.M., Liu, H.Y., Han, H., Zheng, X., Wang, X.G.: Hybrid finite-discrete element modelling of dynamic fracture and resultant fragment casting and muck-piling by rock blast. Comput. Geotech. **81**, 322–345 (2017)

Blair, D.P.: Statistical models for ground vibration and airblast. Fragblast **3**(4), 335–364 (1999)

Blair, D.P.: Non-linear superposition models of blast vibration. Int. J. Rock Mech. Min. Sci. **45**(2), 235–247 (2008)

Blair, D.P.: Limitations of electronic delays for the control of blast vibration and fragmentation. In: Proceedings of the 9th International Symposium on Rock Fragmentation by Blasting, p. 171 (2009)

Carter, C.: Blast design system. In: Third Large Open Pit Mining Conference, Parkille, Aust (1992)

Cho, S.H., Ogata, Y., Kaneko, K.: Strain-rate dependency of the dynamic tensile strength of rock. Int. J. Rock Mech. Min. Sci. **40**(5), 763–777 (2003)

Chung, S.H., Lee, N.H., Hunter, C.J.: A blast design analysis for optimizing productivity at INCO Ltd's Thompson Open Pit. In: Proceeding of 17th Conference on Explosives and Blasting Techniques (1991)

Darling, P.: SME Mining Engineering Handbook (Vol. 1). SME (2011)

Hagan, T.N.: Optimizing the Yield and Distribution of Effective Explosion Energy in Fans and Rings of Blastholes. The AusIMM, Explosives in Mining Workshop, Melboume, Victoria (1987)

Hardy, A.J., Ryan, T.M., Kemeny, J.M.: Block size distribution of in situ rock masses using digital image processing of drill core. Int. J. Rock Mech. Min. Sci. **34**(2), 303–307 (1997)

Harries, G.: The modelling of long cylindrical charges of explosive. In: First International Symposium of Rock Fragmentation by Blasting, Lulea, Sweden (1983)

Harries, G.: Development of a dynamic blasting simulation. In: 3rd International Symposium on Rock Fragmentation by Blasting, Brisbane, Australia (1990)

Holmberg, R.: Design of tunnel perimeter blasthole patterns to prevent rock damage. In Proceeding of IMM Tunnelling'79 Conference, London, pp. 3–6 (1979)

Hudson, J.A., Harrison, J.P.: Engineering Rock Mechanics: an Introduction to the Principles. Elsevier (2000)

Igathinathane, C., Ulusoy, U.: Particle size distribution analysis of ground coal by machine vision ΣVolume approach. In: XXVI International Mineral Processing Congress (IMPC) Proceedings, New Delhi, India, pp. 24–28 (2012)

Jimeno, E.L., Jimino, C.L., Carcedo, A.: Drilling and Blasting of Rocks. CRC Press (1995)

Johansson, D., Ouchterlony, F.: Shock wave interactions in rock blasting: the use of short delays to improve fragmentation in model-scale. Rock Mech. Rock Eng. **46**(1), 1–18 (2013)

Koupouli, N.J., Belem, T., Rivard, P., Effenguet, H.: Direct shear tests on cemented paste backfill–rock wall and cemented paste backfill–backfill interfaces. J. Rock Mech. Geotech. Eng. **8**(4), 472–479 (2016)

Liu, J., Sun, P., Liu, F., Zhao, M.: Design and optimization for bench blast based on Voronoi diagram. Int. J. Rock Mech. Min. Sci. **66**, 30–40 (2014)

Margolin, L.G., Adams, T.F.: Numerical simulation of fracture. In: The 23rd US Symposium on Rock Mechanics (USRMS) (1982)

Morin, M.A., Ficarazzo, F.: Monte Carlo simulation as a tool to predict blasting fragmentation based on the Kuz-Ram model. Comput. Geosci. **32**(3), 352–359 (2006)

Saharan, M.R., Mitri, H.S., Jethwa, J.L.: Rock fracturing by explosive energy: review of state-of-the-art. Fragblast Int. J. Blasting Frag. **10**, 1–2, 61–81 (2006)

Onederra, I.: A fragmentation modelling framework for underground ring blasting applications. Fragblast: Int. J. Blasting Frag. **8**, 3, 177–200 (2004)

Onederra, I., Chitombo, G.: Design methodology for underground ring blasting. Min. Technol. **116**(4), 180–195 (2007)

Palmström, A., Sharma, V.I., Saxena, K.: In-situ Characterization of Rocks. A. A. Balkema Publishers, pp. 1–40 (2001)

Qu, S., Hao, S., Chen, G., Li, B., Bian, G.: The BLAST-CODE model—a computer-aided bench blast design and simulation system. Fragblast **6**(1), 85–103 (2002)

Sanchidrián, J.A., Segarra, P., Ouchterlony, F., López, L.M.: On the accuracy of fragment size measurement by image analysis in combination with some distribution functions. Rock Mech. Rock Eng. **42**(1), 95–116 (2009)

Schill, M., Sjöberg, J.: Finite element simulations of blasting and fragmentation with precise initiation. BLAST & IMPACT (2012)

Scott, A., Onederra, I.A., Chitombo, G.P.: The suitability of conventional geological and geotechnical data for blast design. In: Fragblast 8, pp. 232–238. Editec SA (2006)

Starfield, A.M., Pugliese, J.M.: Compression waves generated in rock by cylindrical explosive charges: a comparison between a computer model and field measurements. Int. J. Rock Mech. Min. Sci. Geomech. Abstr. (1968)

Taylor, L.M., Chen, E.P., Kuszmaul, J.S.: Microcrack-induced damage accumulation in brittle rock under dynamic loading. Comput. Methods Appl. Mech. Eng. **55**(3), 301–320 (1986)

Wang, Q.Z., Li, W., Song, X.L.: A method for testing dynamic tensile strength and elastic modulus of rock materials using SHPB. In: Rock Damage and Fluid Transport, Part I, pp. 1091–1100. Springer (2006)

Wang, M.Z., Shi, X.Z., Zhou, J.: Optimization charge scheme for ring blasting design based on modified Harries' mathematical model and superposition method from a fragmentation perspective: a case study of Tonglvshan copper mine, China. Int. J. Rock Mech. Min. Sci. [Submitted for publication] (2017)

Yi, C., Johansson, D., Nyberg, U., Sjöberg, J.: Numerical simulation for the influence of delay time on the rock fragmentation. In: Proceedings of the 10th International Symposium on Rock Fragmentation by Blasting, Fragblast (2012)

Zhang, Q.B., Zhao, J.: A review of dynamic experimental techniques and mechanical behaviour of rock materials. Rock Mech. Rock Eng. **47**(4), 1411–1478 (2014)

Zhang, Z.X.: Reducing eyebrow break caused by rock blasting in Malmberget mine. In: Proceedings of Thirty Seventh ISEE Annual Conference, SanDiego, USA, 6–9 Feb (2011)

Zhang, Z.X.: Effect of double-primer placement on rock fracture and ore recovery. Int. J. Rock Mech. Min. Sci. **71**, 208–216 (2014)

Zhang, Z.X.: Rock Fracture and Blasting: theory and Applications. Butterworth-Heinemann (2016)

Zou, D.: Theory and Technology of Rock Excavation for Civil Engineering. Springer (2016)

The Impact of Geocell Element Dimensions on Circular Behavior

Soheil Ghareh[1], Sina Kazemian[1(✉)], and Nojan Keshmiri[2]

[1] Faculty of Engineering, Payame Noor University, 19395-4697 Tehran, Iran
Ghareh_Soheil@pnu.ac.ir, kazemian.s@pnum.ac.ir
[2] Faculty of Geotechnic, Islamic Azad University, Tehran Central Branch, Tehran, Iran
nojan2003@yahoo.com

Abstract. The application of auxiliary elements in the development and improvement of soil engineering features has been considered since old times. Nowadays, the efficiency and ability of the soil reinforcement technique in providing practical solutions to different projects has prompted this knowledge to quickly open its place in geotechnical engineering. Considering the impact of the application of modern materials on soil rehabilitation and rehabilitation, geocells are regarded as suitable samples of these materials. The results of experimental studies of the effect of width and height of the geocell element on bearing capacity and P-dimensional settlement have been numerically modeled in this research. It is to be added that laboratory studies in a single-threaded device and analytical studies have been performed using the limited component software ABAQUS 6.11. It can be seen from the results of this study. If the geocell element is used to arrange the soil, the bearing capacity of the beam is increased by 1.65 times compared to the unarmed sample, while the seating capacity in the gantry bearing capacity is proportional to the geocell Not to be armed is just 1.5 times more than that. In addition, when increasing the bearing capacity of the pipelines is of particular importance, it is best to increase the height of the geocell element, but where the size of the site has significant significance, we will have more effective results with the change over the geocell element. An appropriate match between them is considered via contracting and comparing the results of numerical and laboratory studies. In all cases, analytical studies provide more conservative results than the results of laboratory studies.

Keywords: Geocell · Soil improvement · Numerical studies · Laboratory studies · ABAQUS

1 Introduction

Soil, as the most important building materials and the main structural backing, has long been considered by people in construction. However, due to the shear strength, soil Can resist tensile forces, the researchers continuously seek to increase the bearing capacity, resistance and improve its features, and various methods such as mechanical

S. Wang et al. (eds.), *New Solutions for Challenges in Applications of New Materials and Geotechnical Issues*, Sustainable Civil Infrastructures,
https://doi.org/10.1007/978-3-319-95744-9_10

modification, compacting, chemical modification, stabilizing with lime or cement, and the use of the idea of armed soil or the use of auxiliary elements and high tensile strength have been used in this field. In the meantime, the method of soil propulsion is known for its low cost, easy implementation and its great impact on soil features as a suitable method for soil improvement and improvement.

Armed Arms is a structure composed of two different materials, which together minimize their weaknesses and, in this idea, the soil tolerates compressive stresses and armor elements of tensile stresses [1–3]. Nowadays, soil propagation as an effective and reliable method for improving and fixing soil layers is used in addition to increasing load capacity, shear strength and reduction of sedimentation, in fixing surface substrates, road pavement and pavement. In this research, the behavior of the circular motifs based on the sandwich bed with geocell has been investigated numerically and experimentally [4, 5]. Also in laboratory and numerical studies, the effect of parameters of geocell (b) element height and geocell element height (h) on increasing the load bearing capacity and decreasing its seismicity have been investigated. It is worth mentioning that numerical studies have been performed using the limited component software ABAQUS 6.11 [6].

Due to the remarkable progress of technology and the use of modern materials to improve and increase the load bearing capacity of the soil and reduce its settlement, geocells are considered as a suitable sample of these materials. In light of the results of laboratory studies on the behavior of samples arranged by the geocell element, numerical modeling of parameters such as width and height of the geocell element has been investigated in this study.

2 Materials and Method

In order to further investigate the cases affecting the bearing capacity of the joints and their extent and their effect, we first describe how to make the sample, the type of materials used in laboratory studies and how to load the samples, and then each of the parameters tested Such as the geometric element width (b) and the height of the geocell (h) element are separately investigated. The materials used in laboratory studies, including soil and arming elements, are described below.

In laboratory studies, clay was used to create the test bed, and to fill the geochells from the coarse sand and sandy soil. To prepare the preliminary preparations, dry clay is firstly mixed with water. In order to achieve the desired moisture, wet soil is maintained in compressed containers for one week. Then, the soil is placed in a test box monotonously, by placing a wooden plate on the surface and knocking it with a hammer and using the depth specified on the boxes as guides, layers are compacted until the desired height.

2.1 Clay

In this study, the soil used in making the substrate of laboratory samples was silty clay, the characteristics of which are given in Table 1.

Table 1. Physical and mechanical parameters of clay used in manufacturing of laboratory samples

Density of solid beads (GS)	Plastic limit (PL)	Liquid level (LL)	Soil in the USCS system
2.66	17%	40%	CL

2.2 Sand and Gravel

The sand used in this study to fill the geocell elements in a dry laboratory sample, the full specifications of which are given in Tables 2 and 3.

Table 2. Physical parameters of sand used in manufacturing laboratory samples

Minimum porosity ratio (e_{min})	Maximum porosity ratio (e_{max})	Solid grain density (GS)	Effective particle size (D10)	Compression ratio (CC)	Uniformity coefficient (CU)	Soil in the USCS system
0.48	0.66	2.63	360 mm	1.05	2.22	SP

Table 3. Specific gravity and shear strength values of sand and sand in different relative densities of the laboratory sample

The angle of the friction inside the sand (Degree)	Special dry weight (kN/m^3)	Relative density
37	16.4	48%
39	16.6	59%
41	16.8	70%

2.3 Geocell

The archetypal elements of laboratory studies are made of geocell. The geocells are made using polymer axial geogrid with a quadrilateral shape with a pore size of 0.035 × 0.035 m. Geogrid characteristics and joints for geocell construction, obtained from the double standard pressure test in ASTM: D6637, are given in Table 4:

Table 4. Physical parameters of the geogrid and its joints for the construction of the geocell element

Features	Value	
	Geograde	Connection
Ultimate tensile strength	kN/m 20	kN/m 7.5
Failure strain	18%	28%
Primary Module	kN/m 18.3	kN/m 40
Secant modulus at 5% strain	kN/m 160	kN/m 42
Secant modulus at 10% strain	kN/m 143.4	kN/m 29

2.4 Sample

In laboratory and numerical studies, the physical model of the circular shape was made of a solid steel with a diameter of 0.15 m and a thickness of 0.33 m. The soil bed was prepared in a reservoir of length, width and height of 0.9 m. A circular valve with a diameter of 0.095 m was created on the side of the reservoir along an intermediate line at a height of 0.1 m from the bottom of the reservoir. This circular valve is used to create cavities under the bedding clay layers.

2.5 Loading Method

In laboratory studies, cavities have been placed at a constant distance from the surface of the clay layer. Then, on a bed of clay, a sandy layer was placed and was fixed on it by a thin layer of cementitious and epoxy adhesive. Subsequently, a hydraulic jack was loaded in front of the frame reaction, the test steps shown in Fig. 1.

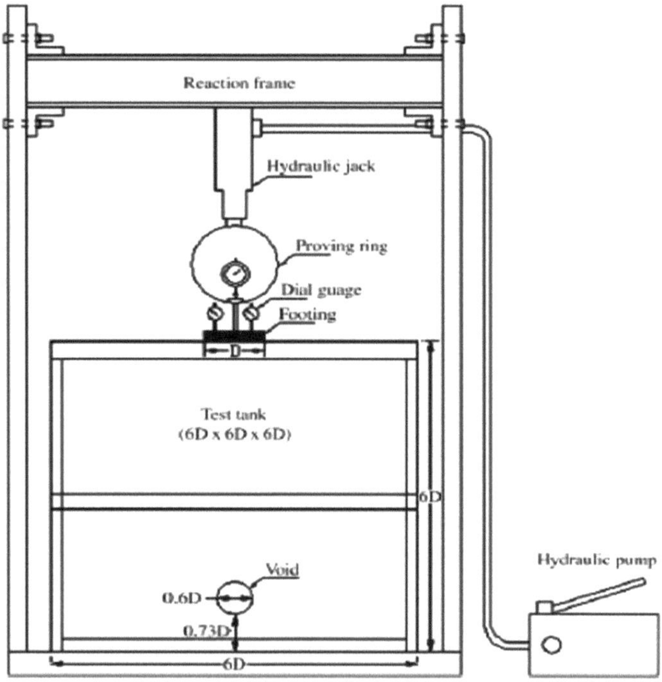

Fig. 1. Laboratory sample loading method

To prepare the test bed, the wet soil is placed in the test box and compressed in thick layers of about 2.5 cm to reach the appropriate height. The degree of saturation (Sr) of the soil is calculated using the specific water content and density, with an average saturation value of about 100%. The geocell layer forms at the beginning of the bed of compressed soils. Due to the superior performance of quadrilateral or diamond-shaped geosets, diamond-shaped elements, all geocell elements are prepared in the form of a quadrilateral or rhombic pattern. After the geocell elements are deployed, these elements are filled with sand and the test is carried out with different variables and the load transmitted to the surface through the graded bars and rings located between the bearings and the loading jacks are evaluated. Path changes are also measured by means of two measuring gauges located on the corner of the beam. It should be noted that in this experiment, the diameter of the P (D), the height of the undrained sand layer (H), the height of the geocell element (h), the geocell (b) width, the measurement gauge (Dg) and the diameter of the hole in Clay bed (dv) is shown in Fig. 2.

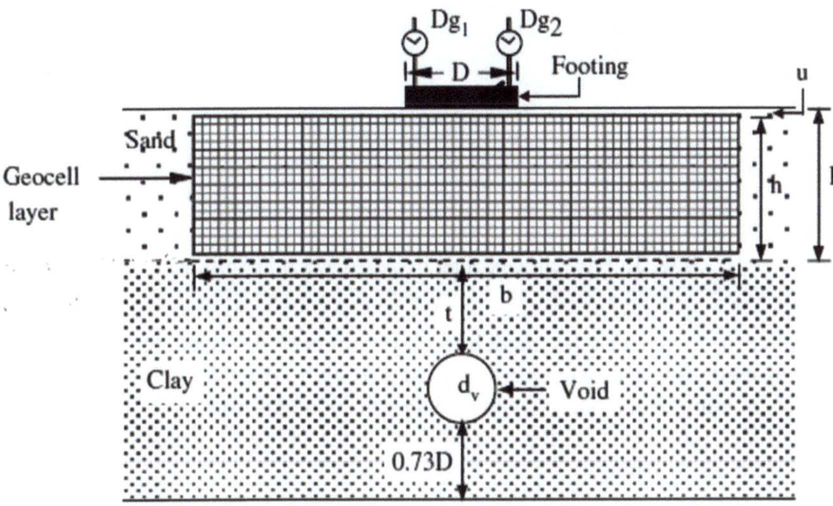

Fig. 2. Test bed preparation method

2.6 Numerical Analysis

In this research, numerical modeling of laboratory samples was performed with the help of ABAQUS 6.11 finite element software. Given that the ABAQUS 6.11 softwares have the ability to solve problems from a simple linear analysis to the most sophisticated nonlinear modeling, it has a very large set of elements for analyzing any geometry This software has been used in the analysis. The next step is to construct the geometry of the model with the dimensions presented in the article, and then define and create all the conditions in the paper, which can be seen in Fig. 3 of the model geometry [6, 7].

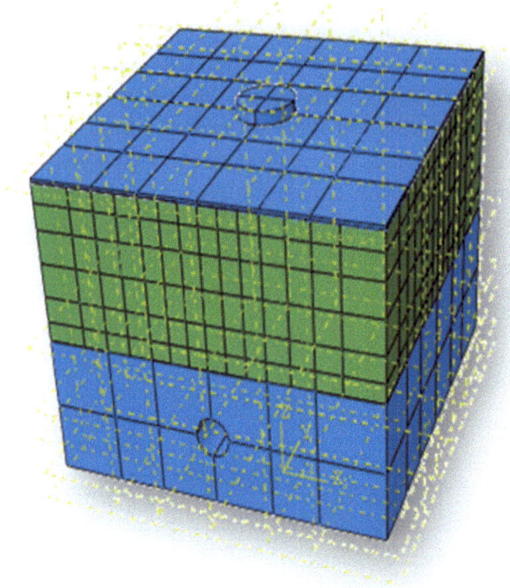

Fig. 3. Program modeling by ABAQUS 6.11 software

It should be noted that for modeling clay nonlinear behavior, the CAM CLAY model was used and sand behavior was simulated with DRUCKER PRAGER model. The results of numerical modeling and also the effects of deformation of P the square of the square and the results of the comparison of the carrying capacity of these two sequences are given below.

3 Results and Discussions

By analyzing the effect of each of the parameters affecting the load carrying capacity and comparing numerical and laboratory studies, the following results are presented:

3.1 The Effect of the Geometric Element Width (B)

The variation in the bearing capacity of the Pei relative to the percentage of its settlement in different latitudes of the geocell (b) is given in Fig. 4, as compared to the diameter of the P-ring (D).

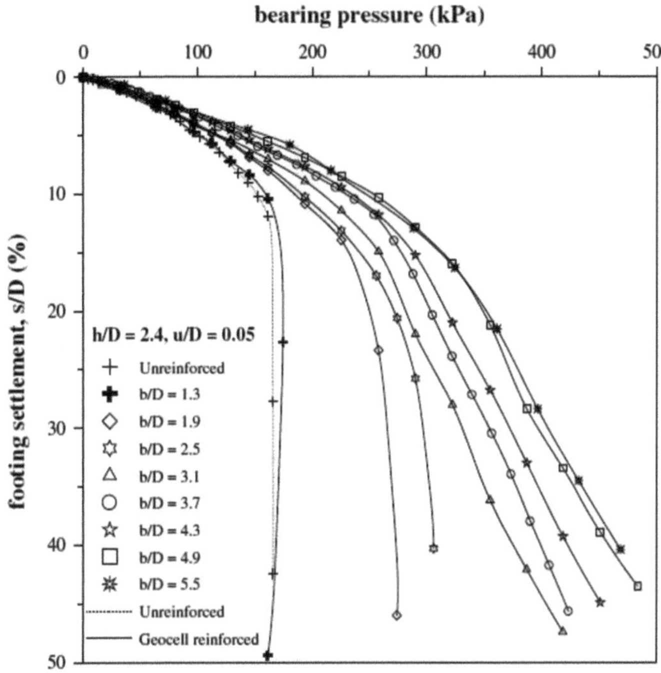

Fig. 4. Changes in the bearing capacity of the Pei relative to the percentage of its settlement in different latitudes of the geocell element (b)

As the results of laboratory and numerical studies show (Table 5), in the case where the width of the geocell element (b) is equal to the diameter of P (D), the presence of the geocell element does not have a significant effect on the improvement of bearing capacity. On the other hand, Table 5 shows that by using a geospatial element with a width close to twice the diameter of the bore (b/D = 1.9), we will see more than 60% increase in bearing capacity and a significant reduction in the summit. The results are agreed well with Sujit et al. (2001) and Sanat et al. (2010) [8, 9].

It is to be noted that the geocell element, which extends over a cavity at a distance of 0.65 dpi (D) and extends beyond the diameter (d = 0.6 D), transfers the pressure to the surrounding soil. It leads to increased performance. Therefore, it can be stated that in order to achieve a more efficient function, the location of the geocell element from the cavities must be at least equal to the diameter (dv), thus increasing the bearing capacity with increasing width of the geocell element (b) Find This increase in performance improvement was significant with increasing b/D to 4.9 1 but since then, given the limited hardness of the geocell material due to its constant height and the size of its elements, the pressure is transferred to the narrower domain. And so there is a slight increase in performance improvement. Also, in order to further investigate the behavior of the pycrays, the physical model in numerical studies using the finite element software ABAQUS 6.11 was used in laboratory studies considering the geocell element width ratio in b/D = 3.7. Results of laboratory studies and the number are given in Fig. 5.

Table 5. The Results of Factor Improvement of Bearing Capacity (IFs) at Different Widths of Geocell Elements (b)

Variable test parameter, b/D	(IFs) Factor Improvement of Bearing Capacity							
	(s/D) = 2%	(s/D) = 3%	(s/D) = 5%	(s/D) = 10%	(s/D) = 15%	(s/D) = 20%	(s/D) = 30%	(s/D) = 40%
1.3	1	1	1	1.03	1.04	1.04	1.04	1
1.9	1	1.21	1.30	1.30	1.43	1.53	1.59	1.63
2.5	1.11	1.24	1.32	1.32	1.49	1.66	1.78	1.84
3.1	1.11	1.38	1.43	1.55	1.59	1.72	1.99	2.25
3.7	1.13	1.43	1.48	1.56	1.71	1.85	2.13	2.40
¾	1.13	1.43	1.53	1.60	1.78	1.94	2.23	2.55
4.9	1.13	1.49	1.66	1.74	1.94	2.13	2.40	2.76
5.5	1.39	1.49	1.73	1.72	1.94	2.14	2.44	2.81

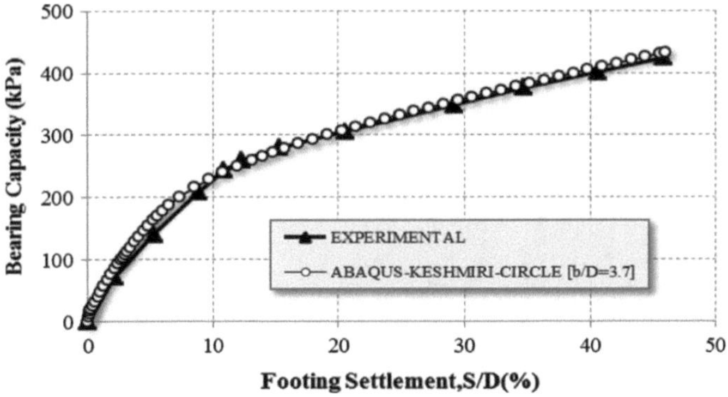

Fig. 5. Modeling of pycrays with geocell with a width ratio b/D = 3.7

As shown in Fig. 5, with the increase in the width of the geocell element (b), the bearing capacity of the beam increases. There is also an appropriate match between the results of laboratory and numerical studies, which indicates the accuracy of the studies carried out.

3.2 Effect of the Height of the Geocell Element (H)

In order to investigate the effect of elevation of the height of the geocentric element, which is formed by the use of a polymer geometric axial geogrid, due to the fact that the diameter of P (D) is constant, we increase the height of the geocell element (h), which is the result of Fig. 6 From the results of laboratory experiments, the bearing capacity of the jet increases with the increase in the height of the geocell element relative to the diameter of the p (h/D).

As shown in Table 6, the bearing capacity of the jet increases with an elevation of the height of the geocell element (h) up to 1.8 times the diameter (D), although in amounts less than 1.8 and in the summits the soil-geocell structure is very high, and subsequently, the geocell element plays a minor role in tolerating the loading of the plies. With increasing height (h), the flexural and shear strength of the geocell element increases and thus can effectively transfer the bridged cavity and push pressure to the enclosed volume of soil. As a result, at a higher altitude of the geocell with respect to the diameter of the pipeline ($h/D \geq 1.8$), the carrying capacity of the pipeline will continue to increase until the P-solution reaches 50%. Finally, we can conclude from this observation that the critical height of the geocell element, which decreases the impact of the cavities on the yield, is about 1.8 times the diameter of the pixel. In order to investigate further the results of laboratory studies, considering the effect of the height of the geocell element (h), a pseudo-modeling with two different heights of the geocell element has been considered. The results are presented in Fig. 7.

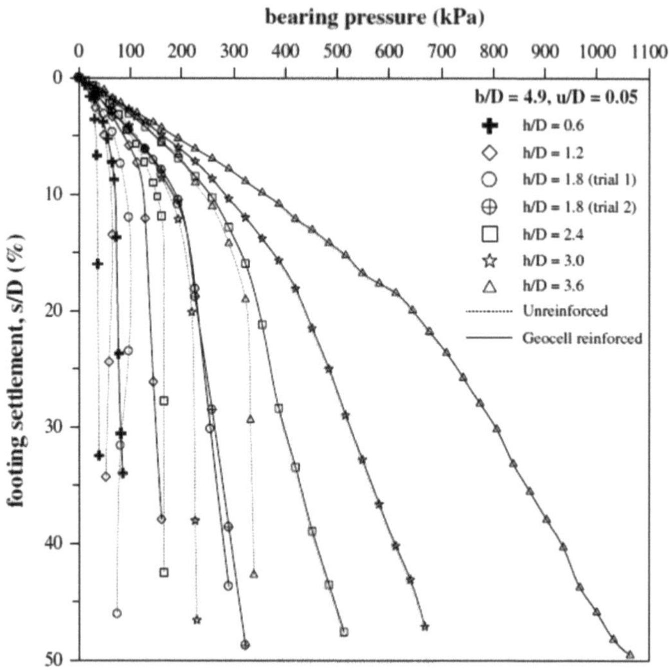

Fig. 6. Changes in the bearing capacity of Pei relative to its summing percentage at different altitudes of the geocell element (h)

As shown in Fig. 7, the process of increasing the carrying capacity of the bearings is clearly evident with the increase in the height of the geocell element. In addition, the numerical modeling results are suitable for experimental results and their difference is about 5%, which indicates the validity of the numerical modeling function in this study. The results are agreed well with Moghaddas et al. (2010), Dash et al. (2012), and Biswas et al. (2013) [10–12].

Table 6. The results of Input Capacity Improvement Factor (IFs) at different altitudes of the geochemistry element (h)

Variable test parameter, h/D	(IFs) Factor Improvement of Bearing Capacity							
	(s/D) = 2%	(s/D) = 3%	(s/D) = 5%	(s/D) = 10%	(s/D) = 15%	(s/D) = 20%	(s/D) = 30%	(s/D) = 40%
0.6	1.31	1.46	1.66	1.98	2.03	2.05	2.11	–
1.2	1.36	1.47	1.66	2.06	2.06	2.14	2.32	–
1.8	1.40	1.50	1.69	2	2.18	2.35	2.59	2.85
2.4	1.13	1.49	1.66	1.74	1.94	2.13	2.40	2.76
3	1.07	1.43	1.48	1.59	1.81	1.99	2.36	2.71
3.6	1	1.10	1.22	1.47	1.72	1.98	2.39	2.73

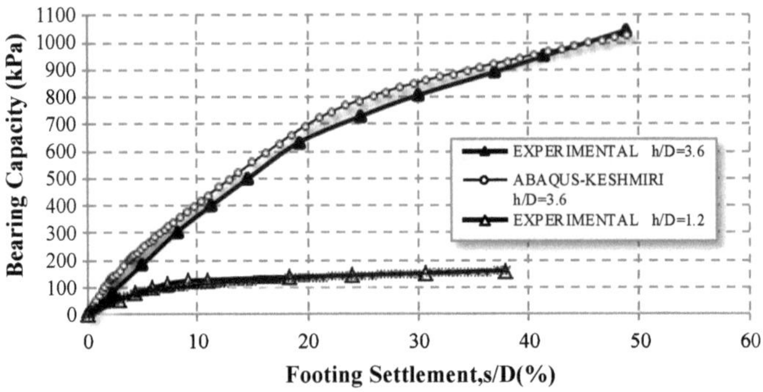

Fig. 7. Comparison of results of geocell-shaped modeling with different heights

4 Conclusions

Subject to the contrast and compare made for the factors affecting the bearing capacity of the joints, such as the geometric element width (b) and the height of the geocell element (h), it can be obtained if the geocell element is used to arrange the soil, as compared to the unarmed specimen, Increases by about 165%, while the percentage of the settlement will only increase by about 15%. Investigating the variations in the width of the geocell element (b) indicates that the increase in this parameter results in a re-distribution of the pressure over a wider area of substrate soil and continuously propagates the pressure P on the cavities and increases the Bearing capacity. It is to be noted that via increasing the height of the geocell element (h), the stability and subsequently the bending and shear strength of the geocell element increase and effectively push the pressure to the volume of the corresponding soil, which increases the bearing capacity of the pile and reduces its settlement.

References

1. Sireesh, S.A., Sitharam, T.G., Dash, S.K.: Bearing capacity of circular footing on geocell–sand mattress overlying clay bed with void. Geotext. Geomemb. **27**(2), 89–98 (2009)
2. Ghiassian, H., Holtz D.H.: Geosynthetic cellular systems (GCS) in coastal application. Report University of Washington, Deptartment of Civil & Environment Engineering, September (2005)
3. Koerner, R.M.: Designing with Geosynthetics. Fifth Edition, Prentice Hall (2005)
4. Heibaum, M.H.: Geosynthetic containers a new field of application with nearly no limits. In: Proceedings of The Seventh International Conference on Geosynthetics 7 ICG-NICE, France, vol. **3**, pp. 1013–1016 (2002)
5. Otto, F.: Tensile Structures. The MIT Press, the Massachusetts Institute of Technology, Boston (1979)
6. Moghaddas Tafreshi S.N., Tavakoli Mehrjardi G.H., Ahmadi M.: Experimental and numerical investigation on circular footing subjected to incremental cyclic loads (2011)
7. Asakereh A., Moghaddas Tafreshi S.N., Ghazavi M.:Strip footing behavior on reinforced sand with void subjected to repeated loading. (2012)

8. Sujit Kumar D., Krishnaswamy N.R., Rajagopal K.: Bearing capacity of strip footings supported on geocell-reinforced sand (2001)
9. Sanat K. Pokharel A., Jie Han A., Dov Leshchinsky B., Robert L. Parsons A., Halahmi I.: Investigation of factors influencing behavior of single geocell-reinforced bases under static loading (2010)
10. Moghaddas Tafreshi S.N., Dawson, A.R.: Comparison of bearing capacity of a strip footing on sand with geocell and with planar forms of geotextile reinforcement (2010)
11. Dash, S.K.: Effect of geocell type on load carrying mechanism of geocell reinforced sand foundations. Int. J. Geomech. **12**, 537–548 (2012)
12. Biswas, A., Murali Krishna, A., Dash, S.K.: Influence of subgrade strength on the performance of geocell-reinforced foundation systems. Geosynth. Int. **20**, 376–388 (2013)

Soil Reinforcement and Slope Stabilization Using Natural Jute Fibres

R. Ramkrishnan[1(✉)], V. Karthik[2], M. R. Sruthy[3], and Animesh Sharma[4]

[1] Department of Civil Engineering, Amrita School of Engineering, Amrita Vishwa Vidyapeetham, Coimbatore, India
r_ramkrishnan@cb.amrita.edu
[2] Department of Civil and Environmental Engineering, University of California at Berkeley, Berkeley, CA 94720, USA
[3] Department of Civil Engineering, Faculty of Engineering and Technology, SRM University, SRM Nagar, Kattankulathur, Kancheepuram, Tamil Nadu 603203, India
[4] Afcons Infrastructure Ltd., Kolkata East West Metro, West Bengal 711103, India

Abstract. This study is aimed at determining the effectiveness of random inclusion of jute fibres on the strength characteristics of two different clayey soils. Fibre length and percentage inclusion in soil mass were taken as the initial study parameters. Unconfined Compressive Strength (UCS) and California Bearing Ratio (CBR) tests were carried out to determine the strength parameters of the reinforced soil. Experimental results were then used to model the reinforced soil as embankment fills in finite element software programs PLAXIS 2D and SLIDE, to determine the safety factors. Results showed that the random distribution of jute fibres had positive effects on both strength parameters and the safety factors. The optimum fibre lengths and their respective percentage inclusions were then established. A mathematical expression, relating the UCS values to the fibre length and percentage inclusion was modelled from the experimental data using linear least square regression method in MATLAB. The data from the FEM software programs were compared with the mathematical expression, and it was observed that the expression held true for various strength parameter values, hence validating the expressions generated.

Keywords: Soil stabilization · Soil reinforcement · Natural fibers
Jute · FEM · Regression

1 Introduction

Soil stabilization can generally be termed as a process of improving the properties of natural soil by physical, chemical or biological means to meet the engineering requirement (Prabhakar et al. 2004; Binici et al. 2005; Sharma and Ramkrishnan 2016; Sahu et al. 2017). Use of fibres can basically be termed as a physical form of soil stabilization. Fibre-reinforced soil can be defined as a soil mass that contains randomly distributed, discrete elements, i.e. fibres. Fibres improve the mechanical behaviour of

S. Wang et al. (eds.), *New Solutions for Challenges in Applications of New Materials and Geotechnical Issues*, Sustainable Civil Infrastructures,
https://doi.org/10.1007/978-3-319-95744-9_11

the soil composite. Tensile resistance in the fibres is mobilized by the normal stresses that act on the soil composite, which imparts greater shear strength to the soil. Most importantly, the behaviour of plant roots is replicated using random discrete flexible fibres, which adds strength to the soil mass and contributes to its stability. Artificial replication of the effect of vegetation has been proved by the usage of fibres for soil mass reinforcement by various laboratory and in situ pilot tests. Use of natural fibres for soil stabilization can be traced back to ancient construction works like Great Wall of China and Babylon (Hejazi et al. 2012; Kumar et al. 2015a, b). Although reinforcement of soil with natural fibres originated in ancient times, it can be concluded that fibre reinforced soil composites have recently gained widespread attention in construction projects. Today, with increasing urbanization and industrialization, the use of soil stabilization techniques is inevitable; hence a systematic study for optimization of the technique is necessary.

With that under consideration, the current study is based on the optimization of fibre reinforced soil stabilization. Of all the available natural fibres, jute fibres are preferred due to its good tensile properties and durability, making it apt for this study. The effect of varying the percentage of jute fibre and the length of the jute fibre was studied by Singh and Bagra (2013), and it was observed that increase in fibre length and percentage weight resulted in an increase in the CBR values. Bairagi et al. (2014) carried out a systematic study on parameters like UCS and CBR in jute fibre-reinforced black cotton soil. An appreciable increase in both the soil parameters was observed. Kumar et al. (2015a, b) studied the effect of the use of jute and coir fibres as soil reinforcement. The length of fibres and diameter of fibres were considered as varying parameters (20, 40, 60, 80 mm). From the results, it was concluded that with an increase in fibre content and fibre length, the CBR value further increased. Results also indicated that the diameter of the fibres also affects the reinforcing effect. Neeraja (2010), and Tapas and Balcshwar (2014), are other significant studies which observed similar results on jute fibre-reinforced soil.

Other natural fibres that have been used for soil reinforcement related studies include coir (Ravishankar and Raghavan 2004; Sivakumar Babu and Vasudevan 2008), palm (Marandi et al. 2008) and sisal (Prabakar and Sridhar 2002; Wu et al. 2014; Mathew and Raneesh 2016; Ramkrishnan et al. 2017). Earlier studies suggest that the degree of improvement of using soil natural fibres depends upon parameters like length of fibre, diameter of fibre and the percentage of fibres (by weight) to the soil mass (Ghavami et al. 1999; Bouhicha et al. 2005; Ahmad et al. 2010; Estabragh et al. 2011). Accordingly, length of the fibre and the percentage inclusion of jute fibres were the considered as varying parameters and their corresponding effects were observed and concluded accordingly.

The present study investigates the strength behaviour and slope stability of the soil reinforced with randomly included jute fibres. In order to understand the effect of inclusion of jute fibres on strength behaviour of clayey soil samples, a series of Unconfined Compressive Strength (UCS) tests and California Bearing Ratio (CBR) tests were conducted. The experimental results obtained from the laboratory tests were then further verified using a mathematical expression which related the strength parameters to the fibre length and percentage inclusion (Ranjan et al. 1996; Sivakumar Babu et al. 2008a, b). A mathematical expression for the reinforced soil strength performance was modelled from the experimental data using linear least square regression method in MATLAB. Furthermore, the practical implications of the use of natural jute fibres were ascertained using numerical modelling. Finite Element Software i.e. PLAXIS 2D and Limit Equilibrium Slope stability software i.e. SLIDE was used for numerical simulation of fibre reinforced slope embankments (Park and Tan 2005; Bhardwaj and Mandal 2006). The results from numerical analyses were then further compared with the mathematical expression, and it was observed that the expression was in good agreement with various strength parameter values.

2 Materials

2.1 Jute Fibres

Jute fibres were collected from Kolkata, (West Bengal, India). To maintain homogeneity of jute fibres properties, they were collected from a single batch of the production.

2.2 Soil Samples

Two soil samples, Soil-B and Soil-R, were collected from Palur and Thirukkazhakundram, Chengalpet (Tamil Nadu, India) respectively. Samples were collected from a depth of 1.5–2.0 metres from the ground level and its properties were tested according to Indian Standard (IS) specifications and are tabulated in Table 1. Both soil samples were classified as clayey soils or fine-grained soils.

Table 1. Properties of Unreinforced soil samples

Sl No.	Parameter		Symbol	Value	
				Soil-R	Soil-B
1	Grain size analysis	Sand	S	13.45%	40.55%
		Silt & clay	M&C	86.55%	59.45%
2	Specific gravity		G	2.25	2.24
3	Liquid limit		WL	51%	35%
4	Plastic limit		WP	19%	16%
5	Shrinkage limit		Ws	8%	13%
6	Plasticity Index		IP	32%	19%
7	Maximum dry density		MDD	15.9 kN/m^3	17.2 kN/m^3
8	Optimum moisture content		OMC	18%	17%
9	Unconfined compressive strength		UCS	171.7 kN/m^2	105.0 kN/m^2
10	California bearing ratio		CBR	6.20%	3.00%
11	Soil classification			CH	CI

3 Methodology

3.1 Experimental Procedure

Previous studies indicated that the shear strength increment was a function of the length of fibres used and the percentage inclusion of fibres. Accordingly, the fibres were cut into different lengths of 0.5 cm, 1 cm and 1.5 cm and were mixed with the soils at different percentage inclusion by soil weight (0.25, 0.5, 0.75, 1, and 1.25%) to reinforce them. The water content during sample preparation was the optimum moisture content of the respective soil type. For each fibre length, Unconfined Compression (UCC) tests were conducted according to IS 2720: Part-10, up to a percentage inclusion at which a drop in the Unconfined Compressive Strength (UCS) value was observed. Thus, the optimum percentage inclusion and fibre lengths in both the soils were established.

CBR tests were conducted on reinforced soil samples in accordance with IS 2720: Part-16. The tests were conducted on jute fibre-reinforced Soil R and Soil B, at optimum percentage inclusion for all fibre lengths (0.5 cm, 1 cm, 1.5 cm), and at optimum fibre length (1 cm) for all percentage inclusions (0.25, 0.5, 0.75, 1, and 1.25%).

3.2 Mathematical Regression

To arrive at a generalized equation that relates the UCS value of the reinforced soil to the fibre length and percentage inclusion, the experimental results were plotted as a 3-dimensional surface using MATLAB. The percentage inclusion and fibre length were chosen as the X and Y axes, while the UCS values were represented along the Z-axis. The variables were then plotted as a 3-dimensional surface using the curve fitting tool. The curves were fit as polynomials with percentage inclusion set to third degree and fibre length set to second degree. Weights were not assigned as both fibre length and percentage inclusion play an equal part in determining the strength improvement. The polynomial equation and the corresponding coefficients for both Soil-B and Soil-R were then obtained from the generated surface.

3.3 Numerical Analyses for Slope Stability

It was observed from the experimental results and mathematical analysis that random inclusion of jute fibres in fine-grained soils increased the strength of soil. In continuation of the analysis, a suitable implication of jute fibre reinforced soil can be high slope embankments which are widely used in the construction of highways, canals, and reservoirs nowadays. For the analyses, Finite Element Software i.e. PLAXIS 2D and Limit Equilibrium Slope Stability Software i.e. SLIDE was used. Slopes of varying heights (10 m, 15 m, 20 m, and 25 m) were simulated in these software programs to investigate the pattern of slope stability. For simplicity, 45° and 60° slopes were considered. In case of Finite element analysis, plane strain condition was considered, and the embankment mesh was kept as fine. In PLAXIS, phi-c reduction technique and for SLIDE, Bishops method was used to ascertain the factor of safety of slope embankments. The results from the experimental tests were used as parameters for numerical simulation of the slope embankments.

4 Results

4.1 UCS Test Results

4.1.1 Effect of Variation of Percentage Inclusion of Jute Fibre on UCS

The UCS variation of jute fibre reinforced Soil-R and Soil-B with the variation of percentage inclusions of jute fibre for each of the fibre lengths studied (0.5 cm, 1 cm, 1.5 cm) are discussed in this sub-head. The stress-strain curves of the UCS tests conducted in Soil-B and Soil-R for 0.5 cm, 1 cm and 1.5 cm fibres are shown in Fig. 1a, b respectively.

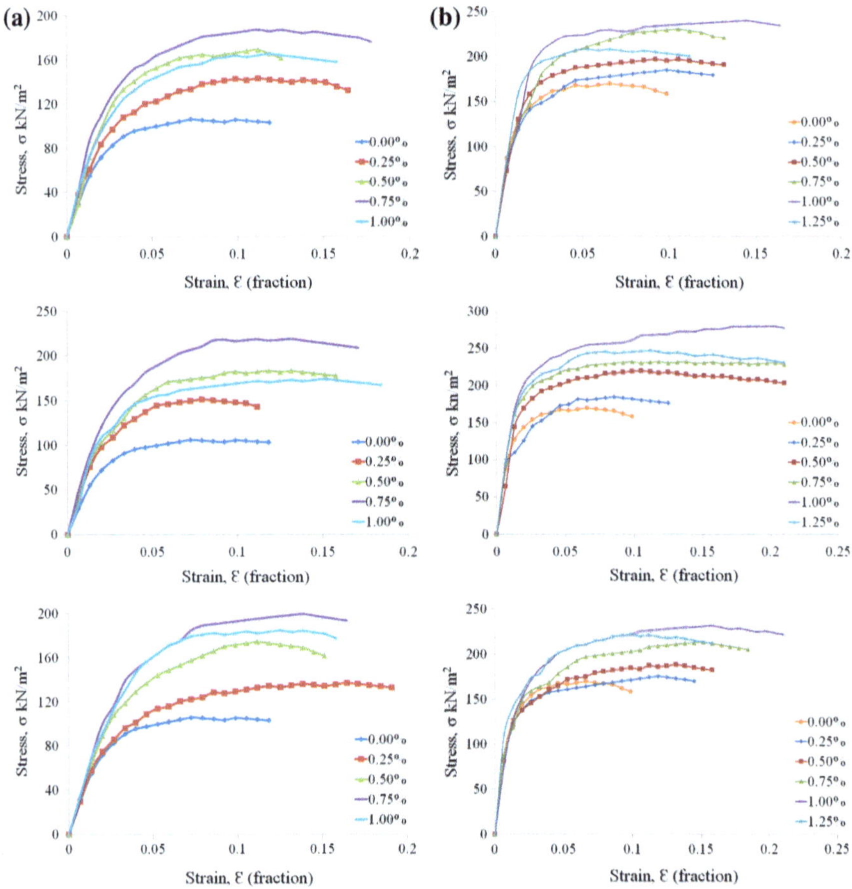

Fig. 1. Effect of varying fibre % on UCS of Jute reinforced (a) soil B and (b) soil R (0.5 cm, 1.0 cm and 1.5 cm)

The increase in percentage inclusion resulted in an increase in UCS irrespective of the fibre lengths used, up to a particular percentage inclusion after which the UCS started decreasing. The percentage inclusion corresponding to maximum UCS was established as the optimum percentage inclusion and its value was 1% in Soil-R and 0.75% in Soil-B for all the fibre lengths.

Soil-R and Soil-B had unreinforced UCS values of 171.7 kN/m^2 and 105.0 kN/m^2 respectively. The addition of 1% of 0.5 cm, 1 cm and 1.5 cm fibres increased the strength of Soil-R by 1.38, 1.55 and 1.32 times respectively whereas the strength of Soil-B was increased by 1.78, 1.93 and 1.87 times on addition of 0.75% of each fibre lengths. These observations confirm that jute fibre was more effective in reinforcing Soil-B than Soil-R. The same can be observed in the Fig. 2, which shows the variation of UCS with percentage inclusion for various lengths of jute fibres in Soil-R and Soil-B. The photographs of tested samples are shown in Fig. 3a, b respectively.

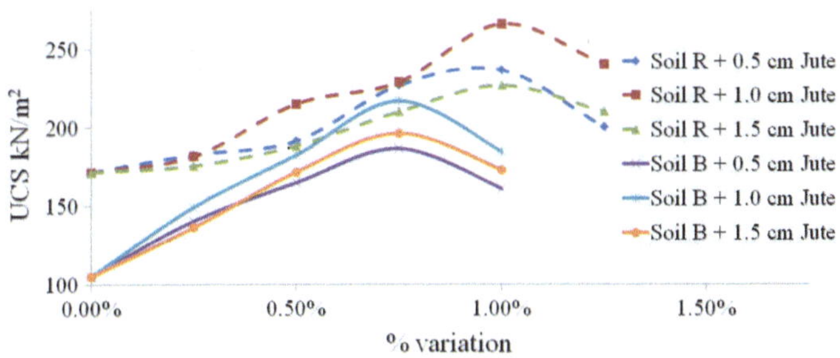

Fig. 2. Variation of UCS with percentage inclusion for various lengths of Jute fibres in Soil-R and Soil-B

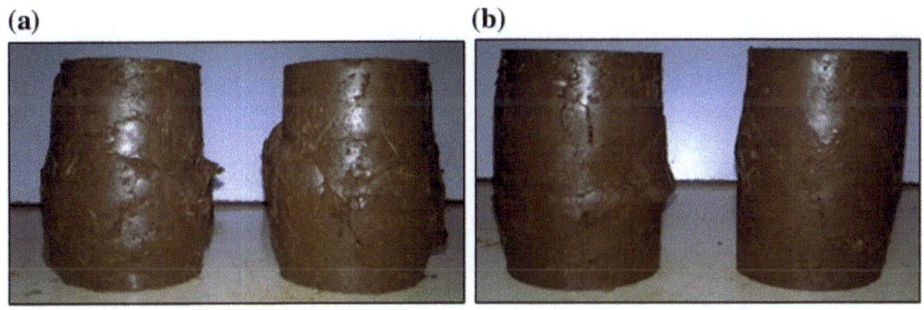

Fig. 3. UCC test samples of (a) Soil R at 1.0% Jute fibre inclusion and, (b) Soil B at 0.75% Jute fibre inclusion

4.1.2 Effect of Variation of Jute Fibre Length on UCS

The UCS variation of jute fibre reinforced Soil-R and Soil-B with the variation of fibre length for all the percentage inclusions are shown in Fig. 4.

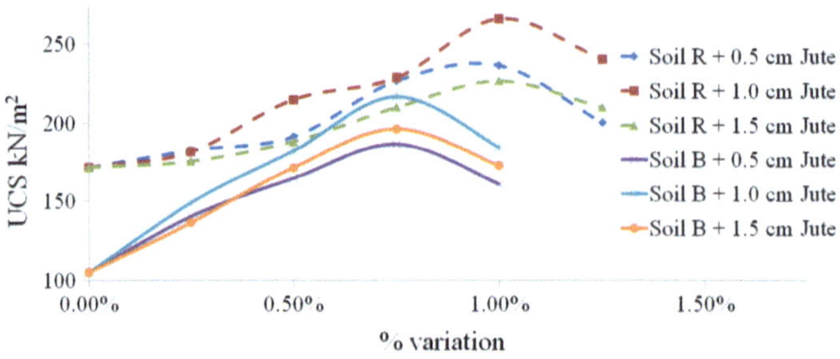

Fig. 4. Effect of Jute fibre length on the UCS of soil

It was found that 1.0 cm fibres, irrespective of the soil type, gave the highest UCS values. Fibres longer than 1.0 cm and higher percentage inclusions caused difficulty in mixing, due to bundling effect. So, the optimum jute fibre length was established as 1 cm for both the soils.

4.2 CBR Test Results

From the CBR tests, it was observed that 1 cm long fibres gave the highest CBR values, irrespective of the soil type or percentage addition. The CBR variation with fibres of various lengths and percentage inclusions are discussed below.

4.2.1 Effect of Jute Fibre Inclusion at Optimum Length on CBR of Soil

Jute fibres of length 1.0 cm were added to the soil samples at varying percentages from 0.25% to 1.25%, with 0.25% increments. The graphs showing the load-penetration curves of the CBR tests conducted in Soil-R and Soil-B are shown in Fig. 5a, b respectively.

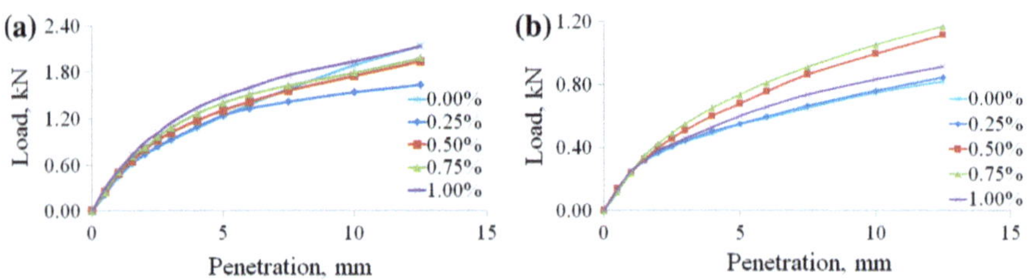

Fig. 5. Effect of varying fibre percentage on CBR of 1.0 cm Jute reinforced (a) soil R and (b) soil B

The addition of 1.0 cm fibres showed a gradual increase in the CBR values till optimum percentage inclusion of 1% in Soil-R and 0.75% in Soil-B, after which the CBR value decreased. An optimum CBR value of 7.55% was obtained in Soil-R and 3.68% in Soil-B, improving it from unreinforced values of 6.2% and 3.0% respectively. An increase of penetration resistance by 1.22 times in Soil-R and 1.23 times in Soil-B was observed. The effect of jute fibre was almost similar in both the soil types.

4.2.2 Effect of Length of Jute Fibre in Soil at Optimum Percentage Inclusion

Jute fibres of varying lengths (0.5 cm, 1.0 cm, and 1.5 cm) were added to the soil samples at optimum percentage addition of 1.0% in Soil-R and at 0.75% in Soil-B. The graphs showing the load-penetration curves of the CBR tests conducted in Soil-R and Soil-B are shown in Fig. 6a, b respectively.

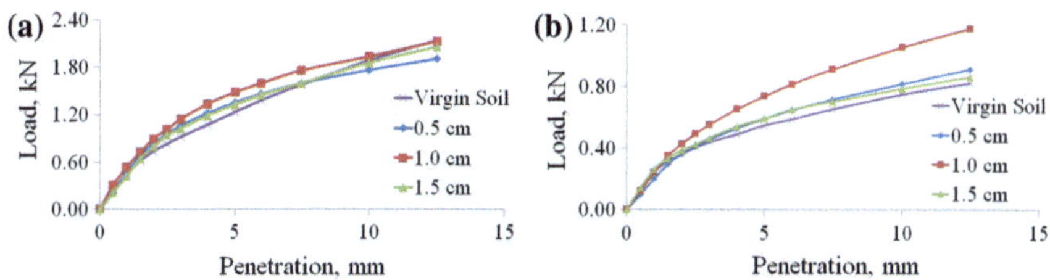

Fig. 6. Effect of varying Jute fibre lengths on CBR of (a) soil R and (b) soil B at optimum percentage inclusion

Fibre lengths of 1.0 cm were found to give higher penetration resistance when compared to 0.5 cm and 1.5 cm fibres at the optimum percentage inclusion. It was observed that at a fibre length of 1 cm (optimum), the CBR value increased with increase in percentage inclusion up to 1% in Soil-R and up to 0.75% in Soil-B, after which it started reducing. CBR variation followed almost the same trend as that of UCS variation. It was also observed that at a percentage inclusion of 1% in Soil-R and 0.75% in Soil-B, the CBR value increased with increase in fibre length up to 1 cm in Soil-R and Soil-B, after which it started reducing. Soil B and Soil R showed a similar pattern in the improvement of penetration resistance which can be easily observed from Fig. 7a, b.

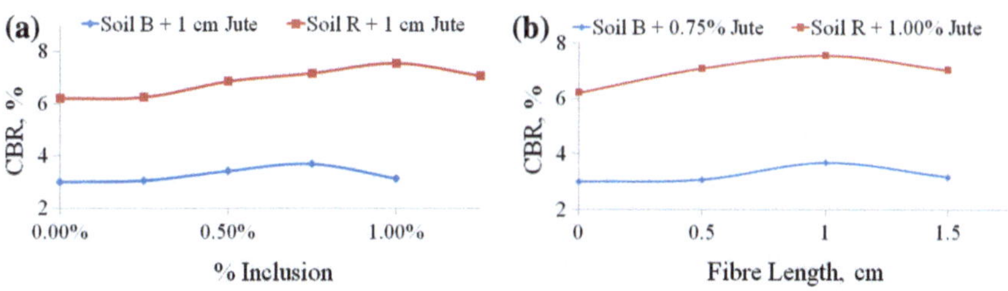

Fig. 7. Comparison of CBR variation with (a) varying percentage inclusion of 1.0 cm Jute fibre (b) varying jute fibre length at optimum inclusion

4.3 Mathematical Modelling Using MATLAB

The UCS values obtained for both Soil-R and Soil-B were used to obtain a mathematical expression relating the UCS values with the fibre length and percentage inclusion using MATLAB. Linear least squares regression method was used to obtain a 3-dimensional surface by plotting the experimental values for Soil-R and Soil-B, as shown in Fig. 8a, b respectively. The above-obtained curves resulted in a sum squared error (SSE) of 332.63 for Soil-R and 92.79 for Soil-B, while the R-square values were 0.9641 for Soil-R and 0.9823 for Soil-B. The polynomial equations of the surfaces for Soil-R and Soil-B are given below in Eqs. (1) and (2) respectively.

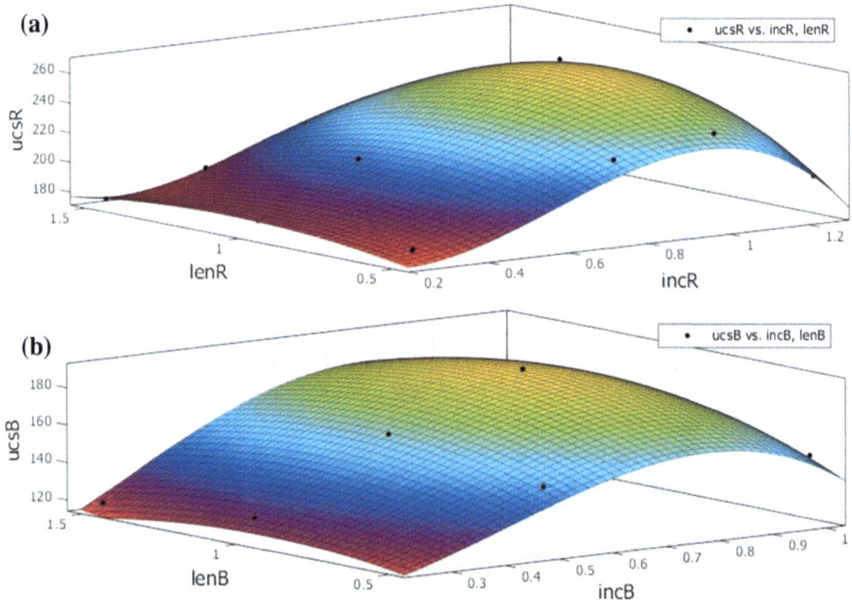

Fig. 8. 3-dimensional surfaces obtained for the mathematical expressions from MATLAB

$$\begin{aligned} q_u = {} & 171.7 - 174.5p + 39.49l + 419.3p^2 \\ & + 112.8pl - 10.35l^2 - 264.6p^3 + 69.91p^2l - 104.3pl^2 \end{aligned} \tag{1}$$

$$\begin{aligned} q_u = {} & 105.0 - 96.60p + 14.00l + 344.7p^2 \\ & + 196.8pl - 18.37l^2 - 257.8p^3 - 70.57p^2l - 48.81pl^2 \end{aligned} \tag{2}$$

where q_u represents the unconfined compression strength (kN/m^2), p represents the percentage inclusion of fibres (%) and l represents the fibre length (cm).

Table 2 shows a comparison between the experimental values and the UCS values obtained from the equations. It can be observed from the table that the values obtained from the equations are similar to the experimental values, with minimal variations. This validates the mathematical equations for the given data set. The size of the experimental data set and the number of soil types used for testing limit the possibility of obtaining a generalised equation for the UCS value of any type of soil that is fibre reinforced with jute fibres. The generalised equation can be attained by conducting the

UCC test on a large number of soil samples which have varying characteristics. Soils under the same classification can be grouped together to obtain a classification specific equation, which would be more relatable than having a single governing equation for all soil characteristics.

Table 2. Comparison of experimental and mathematical UCS values

Percentage inclusion *p* (%)	Fibre Length *l* (cm)	Experimental UCS values (kN/m^2)		Mathematical UCS values (kN/m^2)		Percentage variation (%)	
		Soil-R	Soil-B	Soil-R	Soil-B	Soil-R	Soil-B
0.00	0.0	171.7	105.0	171.7	105.0	0.00	0
0.25	0.5	182.5	117.7	177.1	120.1	2.98	2.05
0.25	1.0	181.5	126.5	185.8	126.6	2.36	0.07
0.25	1.5	175.6	118.7	176.3	117.8	0.39	0.79
0.50	0.5	191.3	151.1	197.3	147.3	3.11	2.49
0.50	1.0	214.8	162.8	207.1	162.6	3.60	0.10
0.50	1.5	188.4	155.0	185.6	156.5	1.47	1.00
0.75	0.5	226.6	159.9	224.6	164.9	0.88	3.12
0.75	1.0	228.6	189.3	239.9	184.6	4.94	2.48
0.75	1.5	209.9	175.6	210.9	176.8	0.47	0.71
1.00	0.5	236.4	151.1	234.3	148.6	0.87	1.64
1.00	1.0	265.9	164.8	259.5	168.4	2.43	2.15
1.00	1.5	226.6	156.0	227.2	154.5	0.28	0.97
1.25	0.5	200.1	–	201.6	–	0.76	–
1.25	1.0	240.3	–	240.9	–	0.26	–
1.25	1.5	209.9	–	209.9	–	0.00	–

4.4 Numerical Modelling of Slope Embankments Using Jute Fibres

Figure 9 shows the variation in the FOS values calculated from PLAXIS and SLIDE for 45° slope. An increment ratio of 1.49 in the cohesion value of Soil-B resulted in an increment ratio of 1.23 and 1.47 in the FOS values of PLAXIS and SLIDE respectively. Similarly, for an increment ratio of 1.80 in the cohesion value of Soil-R, increment ratios of 1.70 and 1.62 was observed in the FOS values of PLAXIS and SLIDE respectively. The cohesion values were taken for the optimum length (1.0 cm) from the experimental results.

Synonymous to the UCS and CBR experimental results, the FOS value gradually increased with an increase in the percentage inclusion. It peaked at 0.75% inclusion for Soil-R and 1.00% for Soil-B, after which there is a decrease in the value. This confirms the optimum percentage inclusions obtained from the experimental results and validates the methodical comparisons between the embankment and the experimental results i.e., random inclusion of fibres in soil with proper compaction in successive lifts till the required height is reached. Also, the FOS value decreases with increase in the

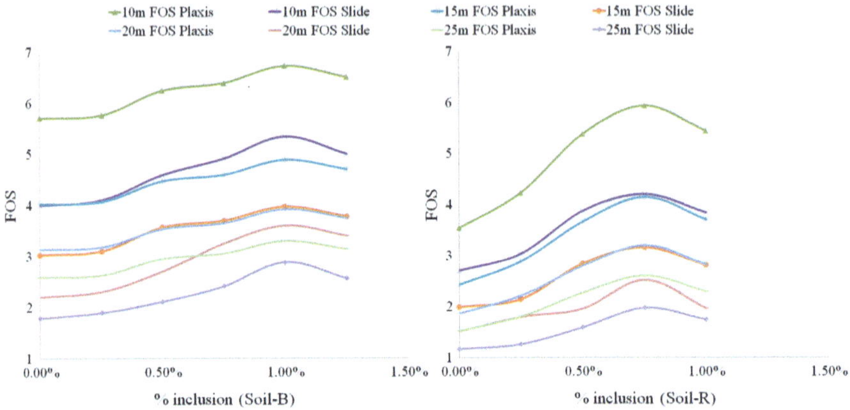

Fig. 9. Effect of varying percentage inclusion and embankment height on FOS of 45° slope

embankment height. This correlates to the fact that the stability of the slope decreases with increase in the slope height.

The variation in the FOS values for 60° slope is depicted in Fig. 10. An increment ratio of 1.49 in the cohesion value of Soil-B resulted in an increment ratio of 1.32 and 1.49 in the FOS values of PLAXIS and SLIDE respectively. Similarly, for an increment ratio of 1.80 in the cohesion value of Soil-R, increment ratios of 1.73 and 1.81 was observed in the FOS values of PLAXIS and SLIDE respectively.

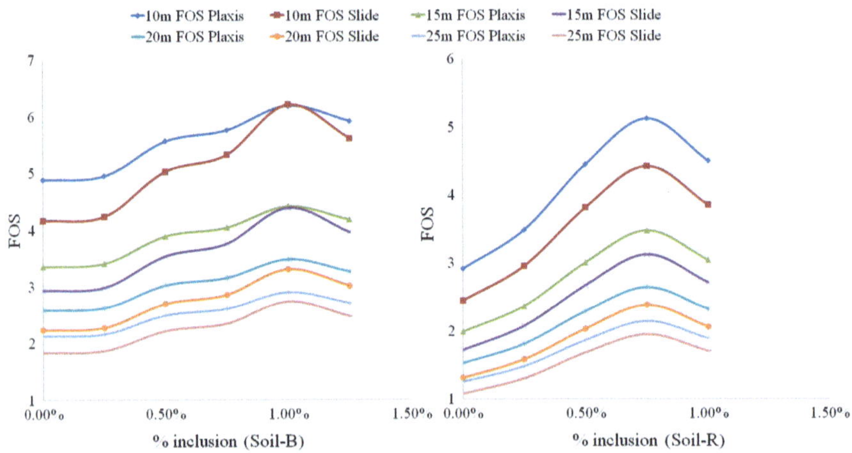

Fig. 10. Effect of varying percentage inclusion and embankment height on FOS for 60° slope

The above graphs and the increment ratios indicate that reinforced soil helps improve the FOS of steeper slopes considerably. It can also be concluded that in case of non-availability of required granular fill for construction of embankments, the available fine-grained soils can be reinforced with sisal fibres to attain a steeper, higher and relatively stable slope embankment. Adopting steeper slopes and reduction of the base width will also be possible using fibre-reinforced embankment fills, which reduces the quantity of material required and the costs involved.

5 Conclusion

An increase in the UCS and CBR values was observed with the addition of jute fibres to both Soil-R and Soil-B, with 1 cm fibre length for both soil types and 1.00% and 0.75% percentage inclusion for Soil-R and Soil-B respectively giving the highest values. Hence, they were reported as the optimum fibre length and optimum percentage inclusion respectively. The variation trends observed for both the soil types were similar as both the types were fine-grained soils with high clay content. The reduction in the shear strength for fibre inclusion beyond the optimum percentage can be attributed to the low specific gravity and low unit weight of jute fibres. The fibres fail to bond with the soil matrix effectively with further addition beyond the optimum values. CBR values of fibre-reinforced soil followed the same trend as that of UCS values and the highest value was observed for 1 cm fibre inclusion. The strength and stiffness of reinforced soil increased with the increase in fibre content which relates to the increase in the CBR value of reinforced soil.

The mathematical equations obtained gave UCS values similar to the experimental values. Hence, these equations can be used to determine the UCS values for various combinations of percentage inclusion and fibre lengths. A possible scope for research in this aspect could be the computation of a generalised equation to obtain the UCS values for various fibre lengths and percentage inclusions. This can be achieved by conducting tests on a large number of soil types with varying parameters. The variations in the basic soil parameters can also be taken into consideration while forming the equation to give more accuracy to the results.

The increment in the strength characteristics of soil indicates a possible application of fibres in the embankments for highway/road projects. Numerical analyses carried out to study the possibility of this application at the optimum fibre length resulted in an increase of the FOS of the embankment until the optimum percentage inclusion and a decrease beyond that. Hence, the stability of the slope increases when it is reinforced with fibres at an optimum percentage addition and length of the fibre. This allows for the economical usage of fibre-reinforced soils in embankments for highway/railway projects. Another practical advantage of using fibres as reinforcements for embankments will be that it will facilitate and accommodate vegetation on facings upon sowing. It will be a crucial aspect for the stability of slopes too, as it will prevent erosion.

References

Ahmad, F., Bateni, F., Azmi, M.: Performance evaluation of silty sand reinforced with fibres. Geotext. Geomembr. **28**(1), 93–99 (2010)

Bairagi, H., Yadav, R., Jain, R.: Effect of jute fibres on engineering properties of lime treated black cotton soil. Int. J. Eng. Res. Technol. **3**(2), 1550–1552 (2014)

Bhardwaj, D., Mandal, J.: Study on polypropylene fiber reinforced fly ash slopes. In: 12th International Conference on Computer Methods and Advances in Geomechanics, Curran Associates, Inc., Red Hook, NY, pp. 3778–3786 (2006)

Binici, H., Aksogan, O., Shah, T.: Investigation of fibre reinforced mud brick as a building material. Constr. Build. Mater. **19**(4), 313–318 (2005)

Bouhicha, M., Aouissi, F., Kenai, S.: Performance of composite soil reinforced with barley straw. Cement Concr. Compos. **27**(5), 617–621 (2005)

Brinkgreve, R., Broere, W., Waterman, D.: PLAXIS 2D. Plaxis B.V, The Netherlands (2004)

Estabragh, A., Bordbar, A., Javadi, A.: Mechanical behavior of a clay soil reinforced with nylon fibers. Geotech. Geol. Eng. **29**(5), 899–908 (2011)

Ghavami, K., Toledo Filho, R., Barbosa, N.: Behaviour of composite soil reinforced with natural fibres. Cement Concr. Compos. **21**(1), 39–48 (1999)

Hejazi, S., Sheikhzadeh, M., Abtahi, S., Zadhoush, A.: A simple review of soil reinforcement by using natural and synthetic fibers. Constr. Build. Mater. **30**, 100–116 (2012)

Kumar, D., Dhane, G., Priyadarshee, A.: Performance of different form of soil reinforcement: a review. Int. J. Sci. Technol. Manage. **4**(1), 667–677 (2015a)

Kumar, D., Nigam, S., Nangia, A., Tiwari, S.: California bearing ratio variations in soil reinforced with natural fibres (a case study Bhopal Bypass Road). Int. J. Emerg. Technol. **6** (2), 95–104 (2015b)

Marandi, S., Bagheripou, M., Rahgozar, R., Zare, H.: Strength and ductility of randomly distributed palm fibers reinforced silty-sand soils. Am. J. Appl. Sci. **5**(3), 209–220 (2008)

Mathew, A., Raneesh, K.: Effect of random inclusion of sisal fibre on strength behaviour of soil. Int. J. Sci. Res. **5**(9), 1702–1707 (2016)

MATLAB and Statistics Toolbox Release 2013b. The MathWorks, Inc., Natick, Massachusetts, United States (2013)

Neeraja, D.: Influence of lime and plastic jute on strength and CBR characteristics of soft clayey (expansive) soil. Glob. J. Res. Eng. **10**(1), 16–24 (2010)

Park, T., Tan, S.: Enhanced performance of reinforced soil walls by the inclusion of short fiber. Geotext. Geomembr. **23**(4), 348–361 (2005)

Prabakar, J., Sridhar, R.: Effect of random inclusion of sisal fibre on strength behaviour of soil. Constr. Build. Mater. **16**(2), 123–131 (2002)

Prabakar, J., Dendorkar, N., Morchhale, R.: Influence of fly ash on strength behavior of typical soils. Constr. Build. Mater. **18**(4), 263–267 (2004)

Ramkrishnan R., Sruthy M.R., Sharma, A., Karthik V.: Effect of random inclusion of sisal fibres on strength behavior and slope stability of fine grained soils. Mater. Today Proc. (in press)

Ranjan, G., Vasan, R., Charan, H.: Probabilistic analysis of randomly distributed fiber-reinforced soil. J. Geotechn. Eng. **122**(6), 419–426 (1996)

Ravishankar, A., Raghavan, K.: Coir stabilised lateritic soil for pavements. In: Indian Geotechnical Conference: Ground Engineering—Emerging Techniques (GREET), Indian Geotechnical Society, Warangal Chapter, Warangal, India, pp. 45–52 (2004)

Sahu, Vaishali, Srivastava, Amit, Misra, Anil Kumar, Sharma, Anil Kumar: Stabilization of fly ash and lime sludge composites: assessment of its performance as base course material. Archiv. Civil Mech. Eng. **17**(3), 475–485 (2017)

Sharma, A., Ramkrishnan, R.: Study on effect of microbial induced calcite precipitates on strength of fine grained soils. Perspect. Sci. **8**, 198–202 (2016)
Singh, H., Bagra, M.: Improvement in CBR value of soil reinforced with jute fiber. Int. J. Innov. Res. Sci. Eng. Technol. **2**(8), 3447–3452 (2013)
Sivakumar Babu, G., Vasudevan, A.: Strength and stiffness response of coir fiber-reinforced tropical soil. J. Mater. Civ. Eng. **20**(9), 571–577 (2008)
Sivakumar Babu, G., Vasudevan, A., Sayida, M.: Use of coir fibers for improving the engineering properties of expansive soils. J. Nat. Fibers **5**(1), 61–75 (2008a)
Sivakumar Babu, G., Vasudevan, A., Haldar, S.: Numerical simulation of fiber-reinforced sand behavior. Geotext. Geomembr. **26**(2), 181–188 (2008b)
Slide. Rocscience, Toronto, ON, Canada (2012)
Tapas, D., Baleshwar, S.: Deformation and strength characteristics of jute geotextile reinforced soils. J. Environ. Res. Develop. **8**(4), 987–995 (2014)
Wu, Y., Li, Y., Niu, B.: Assessment of the mechanical properties of sisal fiber-reinforced silty clay using triaxial shear tests. Sci. World J. **2014**, 1–9 (2014)

Utilization of Geotextile Fabric and Permeable Concrete to Prevent Coastal Erosion

R. Ramkrishnan(✉), N. Narayana Pillai, Azhar Aziz, and R. Gowri

Department of Civil Engineering, Amrita School of Engineering, Amrita Vishwa Vidyapeetham, Coimbatore, India
r_ramkrishnan@cb.amrita.edu

Abstract. In India, the long coastline of South India is deeply connected with the culture, life and traditions of the country. The tourism base of India, especially Kerala, is dependent on the backbone of its exquisite beaches and lush back-waters. Beach conservation, management and development works are executed to aid tourism to a large extent. Nowadays these beaches possess a potential challenge of erosion due to rising and falling of sea levels because global warming and land encroachment towards the beachside. This study focuses on the study of erosion that occurs due to the above-mentioned problems and suggests a novel method to overcome it. The present structures which help in preventing erosion are seawalls, stone revetments, tetra pods etc. All these methods prove to be effective in the short run but the durability and acceptability of the same are not satisfactory. The novel method introduced in this study uses a highly porous geotextile layer laid over a beach area, over which pervious concrete blocks are laid. Water can pass through the revetment easily, while holding the geotextile and underlying soil in place. The blocks are laid in such a way that the waves do not have a direct impact on it, but enter tangentially, thereby reducing the impact and surface drag. After observing the behavior of the revetment exposed to sea waves, it was concluded that laying and anchoring the porous concrete slabs on poly-filter can be an effective armor and a feasible solution for beach protection.

Keywords: Beach erosion · Geosynthetics · Pervious concrete
UFGGBS · Kerala · Wave action

1 Introduction

Kerala has a total coastal length of 580 km of which, about 63% is eroding, confirms the 'Shoreline Change Assessment of Kerala Coast', released by the Union Minister of State for Environment and Forests. About 53% of Kerala's coast now have artificial coast while 10.3% are eroding coastline; experiencing high, medium and low levels of erosion. Only about eight percentage of the 580 km Kerala coastline is stable without erosion or accretion, the study says.

Coastal structures are built to prevent damage of shorelines and also to re-establish the eroded beaches. Seawalls, stone revetments, tetra pods etc. are currently used for coastal protection. Various surveys have shown that, although the present structures help in preventing erosion on the front side, the beaches on the seaward side are eroded

S. Wang et al. (eds.), *New Solutions for Challenges in Applications of New Materials and Geotechnical Issues*, Sustainable Civil Infrastructures,
https://doi.org/10.1007/978-3-319-95744-9_12

and get removed continuously. Seawalls, stone revetments, tetra pods etc. are currently used as erosion protection methods. Seawalls are perpendicular to the waves, reflecting the energy leading to scouring and failure. In stone revetments, erosion rates increase in the front, leading to slope steepening. Gabions consist of rock-filled structures in wire mesh stacked to form a wall. The mesh will eventually corrode leading to collapse (Ali Atef Yousef Masria et al. 2010). Tetra pods endanger beach access and require huge investments for its construction despite absorbing the wave energy (U. S. ARMY CORPS OF ENGINEERS 1981; Zadikoff et al. 1998). These structures settle with time because of continuous wave action, increasing the necessity of reforming the structure over the existing one which makes it expensive. Therefore, coastal structures are imperative to protect the environment, ecology, infrastructures, and economic activities near the shorelines of any populated area (Mitra 2015; Harris and Sample 2009). The rising sea levels and changing shorelines also affect the fishing industry. Thus, for a state like Kerala that depends a lot on agricultural, fishing and tourism sectors, these seawall constructions will not be effective. The current study reports a research work on the use of pervious concrete slabs laid over a geotextile sheet to prevent beach erosion based on a model study carried out by Pillai and Varma (1977). A seminal idea of allowing the waves to enter tangentially over a suitable armor and ride on it safely to the length of the natural run-up and later recede without causing erosion, was put forward. In the case of accretion, the armor will get buried under the sand. In the model study, a thin layer of small stones enclosed in polyethylene nets underlain by a gravel filter was used. The polyethylene nets were found to deteriorate after a few years. It is advantageous to use geo-fabrics of polypropylene for the filter since it will not be exposed to weathering and sunlight and last for longer periods. Considering this idea and allowing the natural movement and percolation of sea water, without facilitating pore pressure generation and sand movement, a combination of pervious concrete slabs and geotextile layer is proposed here as a feasible alternative. If the high velocity waves are deflected and not dissipated at their breaking point near the shore, they cause the beach sand to be washed seaward. The structure is designed to resist the shear at the bottom of the structure. The friction on the slope will also contribute to a smaller extend to the energy dissipation and reduce run up distance. It is a cost-effective alternative, as the materials used are readily available and the maintenance and labor required is comparatively low and no replenishment is required in the longer run.

2 Literature Review

A study by Vera Semeoshenkova and Alice Newton, 2015 on beach quality issues clearly depicts that coastal development and rapid climatic changes are leading causes of beach deterioration. Spatial imbalance in longshore sand transport was concluded as the leading factor of beach erosion in a study done by Seino et al. (2015). Abdulla et al. (2015) studied the behavior of beach sand sediments with monsoonal variations along the Muzhappilangad drive-in beach in Kerala. It showed how erosion brought changes in the wave breaking pattern and highlights the eroding beaches of Kerala with the increase in erosion due to recreational activities, especially drive-ins. Erosion hotspots along the Southwest coast of India was studied by Noujas and Thomas (2015). Ends of seawalls or such structures, improper implementation of erosion preventing structures, fishing gaps, mudbanks etc., were identified as potential erosion spots.

The case study on coastal protection measures done by Ali Atef Yousef Masria et al. (2010), erosion protection methods worldwide were reviewed and divided into four major groups, like sea walls, groins, tetrapods and stone revetments. In sea walls, the face is perpendicular to the waves thus, concentrating the energy and leading to scouring and failure of the wall's foundation and in stone revetments, the erosion rate increases in front of them, leading to slope steepening. Other structures like gabions consist of rock-filled structures in a wire mesh, stacked to form a wall. Even though the wire is plastic coated, the mesh will corrode with time, rupturing the gabion. Structures like groins trap the sediments on one side but increase the erosion on the other side as concluded by Daniel (2001) in their paper on replenishment verses retreat. Thus, the existing structure leads to erosion in a longer run and requires huge investments for its construction and maintenance. Rudlolf et al. (1994) had created a concrete block revetment system for prevention of soil erosion. He proposed a mat of concrete which overlies and holds in place a layer of geotextile which is highly porous, overlying the protected soil area. However, the issue faced here was the requirement of simple geometric shapes and non-uniform wave energy dissipation, which can be overcome using specially designed permeable concrete.

Considering the case study presenting the development of geotextile sand containers, done by Restall et al. (2002), geotextile structures can be effectively utilized to resolve conventional and non-conventional coastal issues. Non-woven sand-filled geotextiles adjusts and adapts readily to varying environments. These structures are user-friendly and ideal for recreational facilities like beaches.

Rodrigo et al. (2012), approved that behavior of concrete structures, mainly from a maintenance point of view, has been excellent and have proved better in resisting fatigue when compared to other offshore structures made of steel or other materials, proving concrete is more apt for offshore structures. Liong and Gunawan (2014) modeled a concrete mattress to assess its maximum deformation with respect to wave height. The mat was flexible and could accommodate deformations and the crack lines due to differential settlement were predetermined along the weaker sections of the mat.

Zhang et al. (2013), suggested that the water-cement ratio has substantial impact on sulphate attack on concrete. Higher the water-cement ratio, greater the expansion and more severe the deterioration of concrete samples, which was confirmed by Weiss et al. (1998), by testing under fatigue loading. By suitable methods, the water content can be reduced to up to 0.3 in the design mix to make the concrete safe against sulphate attack. In-depth studies on effect of mineral admixtures and aggregate sizes on strength, durability and permeability characteristics of pervious concrete was studied by Ramkrishnan et al. (2017). It was concluded from their studies that exempting the use of fine aggregates and using coarse aggregates passing 20 mm and retained on 12.5 mm sieves gave improved strength and permeability to concrete. Durability of this concrete against chemical attacks were improved by the use of admixtures like meta-kaolin and Ultra-Fine Ground Granulated Blast Furnace Slag (UFGGBS).

Investigations on previous literature shows that a strong and highly porous concrete (Draincrete) could be developed which can be used as an armor, underlain by a geo-fabric of polypropylene to successfully address the incessant erosion problem.

3 Materials

3.1 Cement

Ordinary Portland Cement 53 (Chettinad) was chosen owing to its high initial strength and quick setting time. It was sourced from Trivandrum, Kerala. The physical properties as provided by the manufacturer are tabulated in Table 1.

Table 1. Physical properties of OPC53 Cement (Chettinad)

Physical properties	Requirement (ISI2269:2013)	Chettinad OPC 53
Fineness (m^2/g)	225 (min)	285 (min)
Initial setting time (minutes)	30 (min)	161 (min)
Final setting time (minutes)	600 (max)	256 (max)
Soundness (Le Chatelier method) (mm)	10 (max)	0.9
Compressive strength (28 day) (MPa)	53 (min)	58.6

3.2 Coarse Aggregates

Coarse aggregates passing through 20 mm sieve and retained on 12.5 mm sieve was used (Ramkrishnan et al. 2017). It was sourced from a quarry in Trivandrum, Kerala, India.

3.3 Mineral Admixtures

Ultra-Fine Ground Granulated Blast Furnace Slag (UFGGBS)

Ultrafine GGBS, commercially available as Alccofine-1203, is a low calcium silicate based mineral additive which is used as a replacement for silica fumes in high-performance concrete. Its latent hydraulic property and pozzolanic reactivity results in enhanced hydration process (Saurav 2014; Sugapriya et al. 2017). The physical properties of UFGGBS as provided by the manufacturer are tabulated in Table 2.

Table 2. Physical properties of UFGGBFS

Properties	Values
Particle size (microns)	4–6
Specific gravity	2.86
Fineness (cm^2/g)	12000
Bulk density (kg/m^3)	600–700

3.4 Glenium B223

Glenium B233 is a super plasticizer created for applications in high performance concrete where the most astounding performance and toughness is expected. What distinguishes Glenium B233 from other superplasticizers is their unique action mechanism which greatly improves the efficiency of cement dispersion (Ramkrishnan et al. 2017; Ramyalekha and Mallikarjuna Reddy 2017). Its advantages are least labor cost in placing, proven increment in early and ultimate strengths, higher durability and minimal shrinkage and creep, better imperviousness to carbonation and other deteriorating environmental conditions, lesser permeability etc. (He et al. 2017). As the environment where our setup was installed was a sandy beach, the properties like resistance to atmospheric conditions and high strength were of much importance. It was sourced from Coimbatore, Tamil Nadu, India.

3.5 Anchor Rods

Anchor Rods of Standard 0.008 m diameter and 0.5 m length made of Stainless-Steel (SS 404) was used for anchoring the end slabs, ensuring its stability and thereby avoiding overturning due to wave action (Fig. 2). Stainless steel of grade 404 was used, as they have high tensile strength and excellent corrosion resistance owing to the presence of copper and chromium in its composition (Barnaf et al. 2012). Fixities and fasteners of standard size and same material were attached with the anchor rods to increase its stability.

3.6 Anchor Tiles

Anchor rods were attached to the basement with a supporting anchor tile of dimension 0.305 m $\times$ 0.305 m $\times$ 0.02 m. The connection between the anchor rods and anchor tile was done using washers and nuts of standard dimensions.

3.7 Geotextile

Geotextile layer is a thin, adaptable sheet material pervious to water however impenetrable to the substance of the littoral formation (Greiser 1967). It will help in keeping the natural slope intact thus avoiding the slope steepening by sand erosion (Koerner and Te-Yang 1997; Koerner 2000; Koerner and Koerner 2001; Dixon and Jones 2005). The advantages of geotextiles are good permeability with fine filtration, high puncture resistance with high strains to failure, good impact resistance and cushioning ability, high in-plane flow capacity, which concludes the benefits of using geotextile here (Nielsen et al. 2011; Juan Recio et al. 2007; Koerner et al. 1999; Christopher et al. 1997; Giroud 1982). It was laid beneath the pervious concrete block so that sand is retained back and only the water passes through it when the waves retreat. Also, it will allow easy dissipation of any pore pressure developed underneath the structure, without causing any upheaval or deformation to the structure. Installing a geotextile layer was mandatory here as the beach sand was very fine, which occupied the voids of porous concrete blocks, reducing its efficiency over long periods of use. The geofabric layer can effectively protect the porous concrete blocks from clogging and sinking.

Non-woven, non-carbonated polypropylene geotextile of 200GSM, sourced from Pollachi, Tamil Nadu, was used for the study. The advantages of polypropylene are high strain at failure, high resistance to alkalis and detergents etc.

4 Methodology

Wave action and its effects on beaches of Kerala are schematically represented in Fig. 1. Profile 1 in the figure shows the natural profile of the beach before a major erosion takes place. In Profile 2, the cross section of the eroded shore due to continuous action of the waves can be seen. As a solution to this, in the proposed model, the waves are made to tangentially enter a pervious concrete armor layer laid over a geofabric sheet without direct impact, and ride up to its natural run-up, as shown in Profile 3. Turbulence in the breaking wave and higher friction offered to riding accounts for reducing the kinetic energy of the incoming wave. The structure provided is designed to resist the bottom shearing due to the waves, while it moves along the slope.

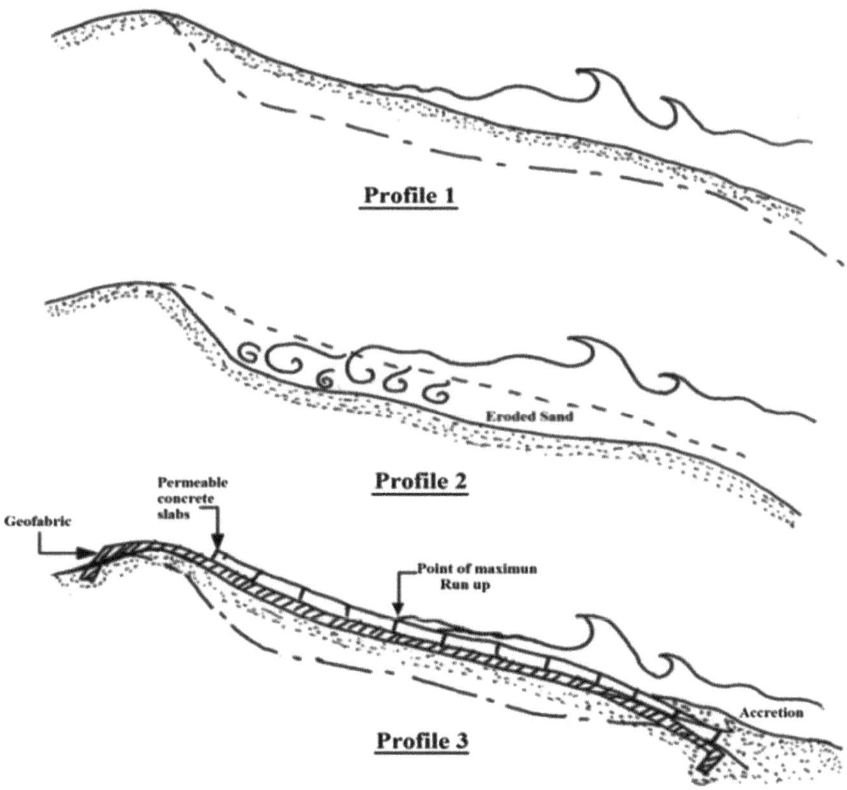

Profile 1 : Shore before erosion
Profile 2 : Shore after erosion
Profile 3 : Shore with permeable concrete slabs and geofabric

Fig. 1. Wave action and its effect on beaches

The pervious concrete slabs were precast and transported to the site after 28 days of curing. To hold the slabs in position, anchorage was provided at an embedment depth based on the survey data, so that the slabs maintain an ascending gradient from the seaside end and the diurnal variation of erosion and accretion does not affect the anchoring. Anchor rods and anchor tiles were embedded in the sand bed after excavating the required design depth (India Patent Application pending, Number 201741031218).

This depth was finalized after close observation of the survey data showing the rate of natural erosion and accretion at the chosen site. The slabs at the edges of the grid were provided with 0.008 m standard anchor rods for ensuring stability. Anchor rods were attached to a supporting tile of dimensions 0.305 m × 0.305 m × 0.02 m, which was underlain and buried underneath the edge slabs at the design depth after excavation. The height of the anchor rods was provided in such a manner that they protrude slightly from the surface of the slabs to enable easy installation of the metal strips which connects the outer slabs. The geotextile sheet was laid over the desired area (5 m × 10 m) by allowing the anchor rods to pass through the geotextile sheet. The pervious concrete slabs were designed and cast with provisions at its center to accommodate the anchor rods and to protrude outwards after placing them over the geotextile. The slabs towards the seaward side were kept embedded in the sand and the designed beach slope was maintained. Stainless steel strips 0.003 m thick and 0.05 m wide were connected to the protruding anchor rods, and used to brace and hold the inner slabs in position to prevent any dislocation or overturning. The ends of the strips were fixed to the anchored edge slabs and the whole setup was exposed to the environment. A schematic diagram showing the cross section of the anchorage setup is shown in Fig. 2.

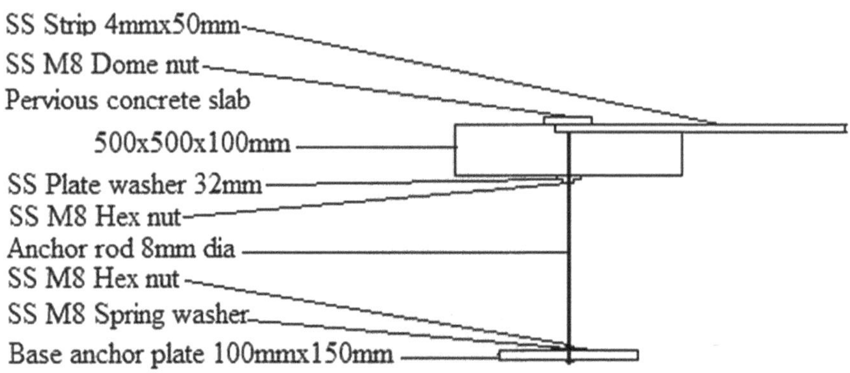

Fig. 2. Schematic diagram of the cross-section anchor setup

4.1 Casting of Pervious Concrete Slabs

Pervious concrete slabs were cast using cement, coarse aggregates, mineral admixtures and water, in the absence of fine aggregates. Pervious concrete, has an interconnected pore structure that freely permits the passage of water to course through (Ravindrarajah et al. 2010). In pervious concrete, aggregate gradation affects void ratio, which influences the strength. From previous studies conducted, it was found that the mix with highest strength and permeability had coarse aggregates of size 0.0125 m and a water cement ratio ranging between 0.27 and 0.30 (Ramkrishnan et al. 2017). As a replacement for silica fumes in high-performance concrete, UFGGBS was used, as it can enhance the hydration process and reduce the setting time (Sugapriya et al. 2017). A total of two hundred slabs of dimensions 0.5 m × 0.5 m × 0.1 m were cast to cover a test area of 5 m × 10 m, as shown in Fig. 3.

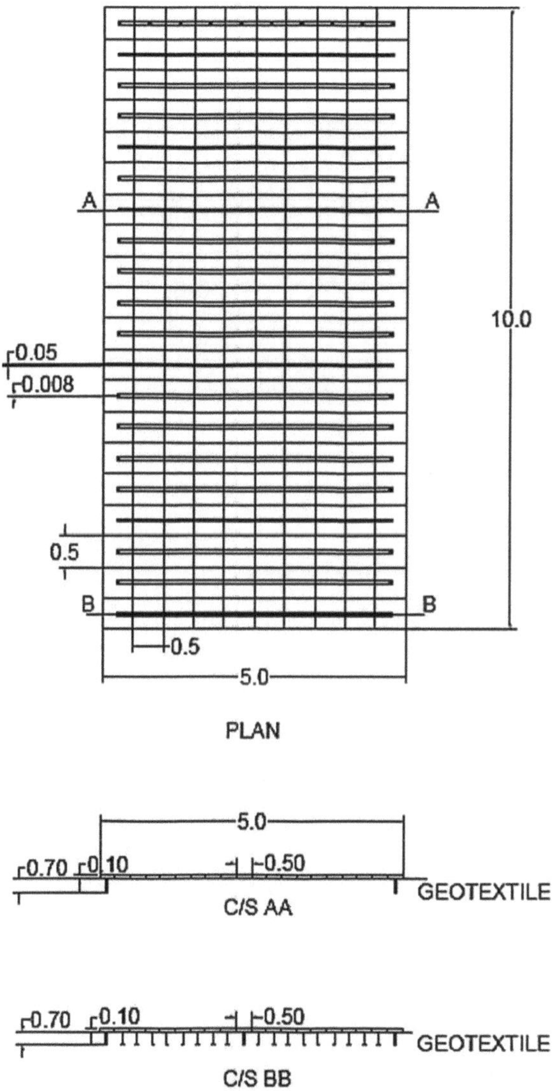

Fig. 3. Plan of the setup implemented (All dimensions are in meters)

Grooves were made in the middle portion of inner slabs to accommodate stainless-steel strips of 0.003 m thickness and 0.05 m width. The pervious concrete slabs during casting is shown in Figs. 4 and 5. The edge slabs were provided with a slanting design at the outer edges to ensure stability and durability against overturning and to prevent scouring of the slabs due to wave impact. Also, prevention of unnecessary turbulence upon wave impact and generation of eddies causing erosion were curbed by allowing smooth and tangential entry of waves over the slanting edges.

The pervious concrete blocks were cast and kept for a curing period of 28 days. Curing was done by immersing the cast slabs completely in a curing tank to avoid any chances of developing shrinkage cracks. The slabs were tested for their strength and permeability properties after the above-mentioned curing period. The properties of the cured slabs after 28 days are tabulated in Table 3.

Fig. 4. Pervious concrete inside formwork

Fig. 5. Slab after removing the formwork

Table 3. Properties of pervious concrete slabs

Properties	Values
Compressive strength	16.63 MPa
Porosity	24.84%
Density	1931.25 kg/m^3
Coefficient of permeability	1.05 cm/s

4.2 Work Prior to Laying

The site which was selected for the study is shown in Fig. 6. The average wave height at the site was found to be about 1.2 m in the season the experiments were conducted. Wave period was observed to be nearly 9 s. This was found to be in order with previous observations made by Asharaf et al. (2001). There is a breakwater structure on one side of the site and there are no other structures near to the selected area. The erosion/accretion levels at three different points were measured from three different points A, B and C as shown in Fig. 7 using a theodolite. The stations were selected in such a way that they are collinear with an interval of 7.5 m and the line is approximately parallel to the beach. This line was fixed as the datum. Point O, as shown in Fig. 7 was marked and established as the control point. Using the basic survey principles, the Reduced Levels (R.L.) were taken from a levelling staff held at the required positions, A, A', A", B, B', B", C, C' and C", which were fixed such that they lie along 3 lines perpendicular to the beach as shown in Fig. 7. The R.L.'s were calculated with respect to the nearby road level, which was considered as the zero level. As diurnal variation was observed at these levels showing alternating accretion and erosion processes, the interval at which the readings were noted was fixed as 6 h between 6 am and 6 pm daily. This was done so that the embedment depth required for the edge slabs and its anchoring could be decided and the amount of accretion and erosion of sand on the beach could be quantified. The schematic diagram showing the average diurnal erosion rates at the central portion of the proposed revetment, A, A', and A" is shown in Fig. 8.

Fig. 6. Site for implementation

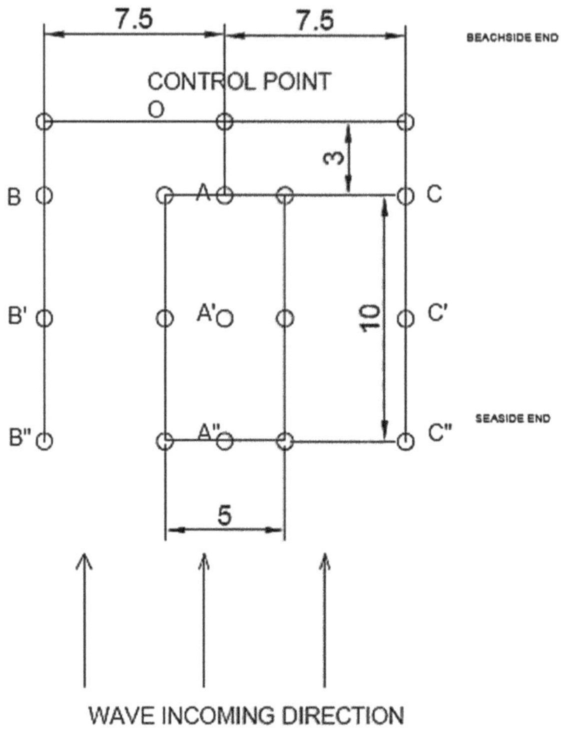

Fig. 7. Slab grid and station points

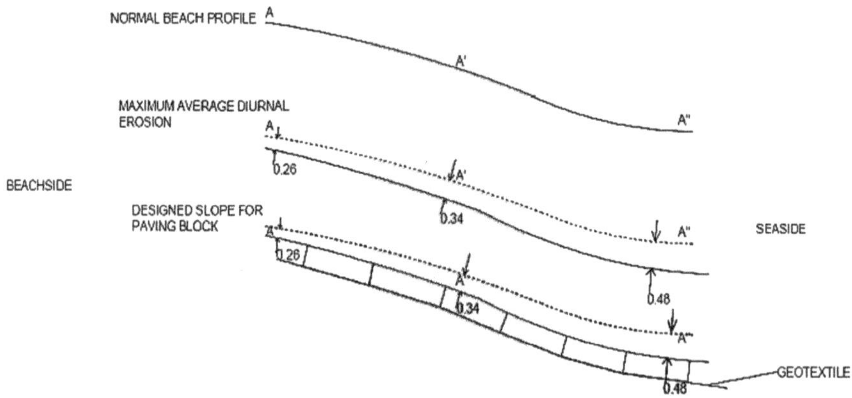

Fig. 8. Schematic diagram with average erosion rates, showing proposed depth of embedment of slabs

The cross section of normal beach profile is represented by the first curve shown in Fig. 8. The second curve in the figure depicts the maximum average diurnal erosion at various points along the beach (A, A', A", B, B', B" and C, C', C"). The erosion at these points were noted so that the depth of embedment can be designed based on the alternating erosion or accretion at various points. The results showed a maximum erosion of 0.48 m at the seaside end and a minimum of 0.26 m at the beach side. Therefore, it was decided to install the anchor tiles at an embedment depth greater than this depth to safely hold the concrete slabs in position. A depth of 0.3 m was adopted at the beachside and a depth of 0.6 m was adopted at the seaside.

4.3 Laying of Slabs

The site during implementation of the structure is shown in Fig. 9a and b. A total of 200 slabs of dimension $0.5 \times 0.5 \times 0.1$ m had to be laid on an area of 5 m $\times$ 10 m at the beach. The beach sand was excavated along the outer edge of the required area. Anchor rods were attached to a supporting anchor tile of size $0.305 \times 0.305 \times 0.02$ m at the bottom and were placed in the excavated area before backfilling the soil. The steel strips connected to the anchor rods protruding from the edge slabs to prevent the overturning of inner slabs can be observed from the Fig. 9b.

To allow the waves to enter the revetment tangentially, the first few rows of slabs towards the seaside (slabs first hit by the waves) were embedded in the sand and the remaining rows were laid at a gradually increasing slope. The natural slope of the beach was maintained as closely as possible during construction of the revetment.

5 Results and Discussion

On the next day, it was found that, towards the seaside end there was slight scouring underneath the edge slabs closer to the breakwater structure and accretion on the opposite edge. During the diurnal accretion period, the slabs were partially covered with sand on one side, extending up towards the beachside, showing good amount of accretion on the revetment, as some waves entered the revetment obliquely from that side, as shown in Fig. 10.

During the erosion phase, these slabs were exposed and in place, but the slabs on the other side of the revetment, closer to the breakwater structure were found to have slightly displaced from its position due to scouring caused by the frontal waves. The waves started progressively scouring the edge slabs and worked its way inwards, eroding the sand on one side of the revetment. The process continued till the erosion phase ended and caused displacement of a few slabs from the edge. The frontal waves riding on the revetment during the erosion phase is shown in Fig. 11a and its result mentioned above can be observed from the Fig. 11b.

It can be clearly stated here that this happened as the study area was minimal in width, and will be avoided when the revetment is extended over a larger width along the beach and/or the slabs towards the edges are embedded towards the sides or terminated on an abutment or a wall. However, the slabs towards the centre of the revetment were stable and didn't show any noticeable movement or settlement, even

(a) Setup during installation

(b) Setup during installation

Fig. 9. (a) Setup during installation. (b) Setup during installation

Fig. 10. Accretion on the setup

(a) Revetment exposed to frontal sea waves

(b) Revetment exposed to frontal sea waves

Fig. 11. (a) Revetment exposed to frontal sea waves. (b) Revetment exposed to frontal sea waves

after a considerable period of observation, allowing the authors to conclude that the sand underneath the centre portion of the revetment was untouched due to the wave action.

On the other hand, during the accretion phase, the waves continued to ride over the revetment, frontal waves causing minimal disturbances and along with the oblique waves, depositing good amount of soil above the revetment. The structure was found to finally merge with the natural beach profile during periods of accretion as shown in Fig. 10.

It was observed that as the oblique wave front was moving on the pavement, it was moving on it safely without any damage to it. When the waves were normal to the beach, the waves entered the pavement tangentially and went up riding on it as shown in Fig. 11a. Then it receded as shown in Fig. 11b, proving that the pavement is safe from the wave action. However, the experimental pavement could be made only for a narrow width of 5 m which was not sufficient enough to bury the blocks on the sides to the scour depth and provide a gradual slope inward. The waves began dislocating the side slabs and the damage worked towards the center of the pavement after some time. This situation could be avoided if the blocks on the sides are well buried under the sand below the possible scour depth or the pavement terminated in an abutment or a wall.

6 Conclusions

The high permeability of the slabs and geotextile allowed natural percolation of water and smooth dissipation of any pore pressure developed beneath it. The protection was given on an experimental basis with a minimal width of 5 m. In a limited time of study, the wave action on the central portion of the pavement was observed to assess whether the wave could ride on and later recede from the pavement without damage. Though the edge slabs at the seaside end near the existing breakwater was affected due to the frontal waves during erosion periods, the sand below the central region of the revetment was completely intact for a considerable time. Hence it can be concluded that the proposed idea, can efficiently prevent beach erosion, when implemented for longer distances or when the slabs towards the edges are embedded, maintaining a gradual slope inward, or by terminating the revetment on an abutment or wall. The study couldn't be carried out on a much wider pavement due to the limitation of funds and time, however, in the future, the experiments can be replicated over a wider area, with continuous monitoring of the performance of the setup. Scour size and depth if any, and any settlement or displacement of the pavement can be assessed, quantified and evaluated for further developing the idea. From this study, it is clear that a revetment having a Draincrete layer Underlain by a Poly-Filter (DUPF) layer can be effectively used as a feasible solution to the problem of beach erosion (Patent Pending, Application Number 201741031218). Further studies and site-specific modifications can be made on the structure and in-depth studies on the topic can be carried out to provide specific embedment depths, anchorages, bracing methodologies, revetment gradient and abutment designs.

Acknowledgments. We would like to express our gratitude to the Water Resources and Irrigation Division of Government of Kerala and Amrita Vishwa Vidyapeetham, India for funding the project and for providing the necessary facilities required for the successful completion of the study. We express our heartfelt thanks to Dr. M. C. Dathan, Scientific Advisor to the Chief Minister of Kerala, and Dr. Sasangan Ramanathan, Dean, Amrita School of Engineering for their constant support and encouragement. We would also like to thank Dr. K. M. Mini, Chairperson, and Mr. Srinivasan K., Department of Civil Engineering, Amrita School of Engineering, for their support and guidance.

References

Abdulla, P.K., et al.: Changes to a drive-in beach in southwest india due to severe erosion and subsequent reformation. Aquatic Procedia **4**, 357–364 (2015)

Asharaf, T.T.M, Nair, R.P., Sanjana, M.C., Muraleedharan, G., Kurup, P.G.: Surface wave statistics and spectra for Valiathura coastlines, SW coast of India. Indian J. Mar. Sci. **30**, 9–17 (2001)

Barnaf, J., Bajer, M., Vyhnankova, M.: Bond strength of chemical anchor in high-strength concrete. Procedia Engineering **40**, 38–43 (2012)

Cohen, J.E., et al.: Estimates of coastal populations. Science **278** (5341), 1209–1213 (1997)

Daniel, H.: Replenishment versus retreat: the cost of maintaining Delaware's beaches. Ocean Coast. Manag. **44**(1), 87–104 (2001)

Dixon, N., Jones, D.R.V.: Engineering properties of municipal solid waste. Geotext. Geomembr. **23**(3):205–233 (2005)

Douglas, G.: KOERNER, L. Linnaeus: nature and nation. Harvard University Press, Cambridge, Massachusetts: 1999. Pp viii, 298; 4 figures. Price£ 27.50. ISBN 0-674-09745-9, pp 281–282 (2001)

Fernandez, R.P., Pardo, M.L.: Offshore concrete structures. Ocean Eng. **58**, 304–316 (2013)

Giroud, J.P.: Design of geotextiles associated with geomembranes. In: Proceedings of 2nd International Conference on Geotextiles, vol. 1. Industrial Fabrics Association Int., Roseville, MN, (1982)

Harris, L.E., Sample, J.W.: The evolution of multi-celled sand-filled geosynthetic systems for coastal protection and surfing enhancement. Reef J. **1**(1), 1–5 (2009)

He, Z., Li, L., Du, S.: Mechanical properties, drying shrinkage, and creep of concrete containing lithium slag. Constr. Build. Mater. **147**, 296–304 (2017)

Koerner, R.M., Te-Yang, S.: The evolution of geosynthetics. Civ. Eng. **67**(7), 62 (1997)

Koerner, R.M.: Emerging and future developments of selected geosynthetic applications. J Geotech. Geoenviron. Eng. **126**(4), 293–306 (2000)

Koerner, R.M., Koerner, G.R.: Geosynthetics design beyond factor of safety: risk assessment using probability of failure analysis. In: Proceedings GRI-15, Hot Topics in Geosynthetics-II, pp 235–253 (2001)

Liong, G.T., Gunawan, A.: Modeling of concrete mattress for shore protection by plax is 2D. Int. J. Appl. Eng. Res. **9**(21), 9715–9728 (2014)

Greiser, M.R.: Prevention of beach erosion and encouragement of land restoration. No. US3344609 A, 1967

Masria, A., et al.: Coastal zone issues: a case study (Egypt). Procedia Eng. **70**, 1102–1111 (2014)

Mitra, A.: Application of geotextiles in coastal protection and coastal engineering works. Int. Res. J. Environ. Sci. **4**(4), 96–103 (2015)

Nielsen, A.F., Mostyn, G.: Considerations in applying geotextiles to coastal revetments. In: Proceedings of Symposium Coastal and Marine Geotechnics, Aust. Geomechanics Society and NSW Maritime Panel, Engineers Australia, October. pp 91–102 (2011)

Noujas, V., Thomas, K.V.: Erosion hotspots along southwest coast of India. Aquatic Procedia **4**, 548–555 (2015)

Pillai, N.N., Verma, D.V.S.: Shore protection using stones enclosed in nets. Coast. Eng. **1**, 349–358 (1977)

Ramkrishnan, R., Pillai, N.N.: A system and a method for preventing beach erosion. India Patent Pending (2017), Application Number 201741031218

Ramkrishnan, R., Abilash, B., Trivedi, M., Varsha, P., Varun, P., Vishanth, S. Effect of mineral admixtures on pervious concrete. In: Materials Today Proceedings, International Conference on Advances in Materials and Manufacturing Applications, Bengaluru, August 2017

Ramyalekha, B., Mallikarjuna Reddy, J.: Study on properties of fresh and hardened self compacting concrete with varied percentages of metakaolin as mineral admixture (M40 Grade). IOSR J. Mech. Civ. Eng. **14**(1), 15–21 (2017)

Recio, J., Oumeraci, H.: Effect of deformations on the hydraulic stability of coastal structures made of geotextile sand containers. Geotext. Geomembr. **25**(4), 278–292 (2007)

Restall, S.J., et al.: Case studies showing the growth and development of geotextile sand containers: an Australian perspective. Geotext. Geomembr. **20**(5), 321–342 (2002)

Saurav, A.K.G.: Experimental study of strength relationship of concrete cube and concrete cylinder using ultrafine slag Alccofine. Int. J. Sci. Eng. Res. **5**(5), 102 (2014)

Seino, S., et al. Essential aspects of beach erosion-Lessons from devastation of Ichinomiya coast, Japan. Procedia Engineering **116**, 446–453 (2015)

Semeoshenkova, V., Newton, A.: Overview of erosion and beach quality issues in three Southern European countries: Portugal, Spain and Italy. Ocean Coast. Manag. **118**, 12–21 (2015)
Stine, J.K., Michael, C.R.: The US Army Corps of Engineers and Environmental Issues in the Twentieth Century: A Bibliography. DIANE Publishing, (1984)
Sugapriya, P., Ramkrishnan, R.: Crumb rubber recycling in enhancing damping properties of concrete. In: IOP Conference Series: Materials Science and Engineering, International Conference on Advances in Materials and Manufacturing Applications. Bengaluru, August 2017
Sugapriya, P., Ramkrishnan, R., Saravanamurugan, S., Keerthana, G.: Experimental investigation on damping property of coarse aggregate replaced rubber concrete. In: Materials Today Proceedings, International Conference on Advances in Materials and Manufacturing Applications. Bengaluru, August 2017
Weiss, W.J., Yang, W., Shah, S.P.: Shrinkage cracking of restrained concrete slabs. J Eng. Mech. **124**(7), 765–774 (1998)
Zadikoff, G.P.E., et al.: Concrete tetrahedrons and sand-filled geotextile containers: new technologies for shoreline stabilization and other uses. J Coastal Res. 261–268 (1998)
Zhang, M., et al.: Study on the expansion of concrete under attack of sulfate and sulfate–chloride ions. Constr. Build. Mater **39**, 26–32, (2013)

Assessment of Mass Movements and Critical Phreatic Levels in Soil Slopes

D. Ravichandran[1], E. Nishok Kumar[1], R. Ramkrishnan[1(✉)], Karthik Viswanathan[2], S. Sandeep[1], and K. Manasa[1]

[1] Department of Civil Engineering, Amrita School of Engineering, Amrita Vishwa Vidyapeetham, Coimbatore, India
r_ramkrishnan@cb.amrita.edu
[2] Department of Civil and Environmental Engineering, University of California at Berkeley, Berkeley, CA, USA

Abstract. Tectonic movements and vibrations of the earth cannot be controlled, which causes devastating natural hazards like landslides and earthquakes which have accounted for many lives in the previous years. The major reasons for landslides are heavy rainfall, liquefaction, rise in pore water pressure, floods, etc. This experimental study focuses on identifying the Critical Phreatic Level (CPL) of different soil types for different slope geometries. Different soil types were modeled in a tank of dimensions 2.30 × 1.00 × 1.25 m to simulate the natural field conditions like field density, ground water flow and slope angle in the laboratory with scaled down slopes of specific angles, based on the natural angle of repose of the soil. Density closely resembling the natural field density was obtained by air pluviation and a constant water inflow from an adjacent chamber was provided to simulate groundwater flow. The slope geometry was modeled, initial conditions were set and the phreatic level in the slope was continuously monitored until the slope fails with considerable slope displacement. The soil properties such as permeability, bulk unit weight, specific gravity and angle of repose obtained from laboratory tests were used as input parameters to model the slopes in PLAXIS 2D. The displacement values obtained from the software were compared with the displacement values obtained from the experiment, and were found to be similar, thereby validating the results.

1 Introduction

Slope of a soil mass is the gradient or angle of inclination of the soil surface from the horizontal, with one end at a higher level and another at a lower level, also called rising or falling of the earth surface. Formation of these sloped surfaces gives rise to different topographical features on the earth surface, which is either natural or man-made formations with varying slope geometries. These slopes collapse when adverse conditions act upon it. Stability of the soil is its potential to withstand and restrict movement. Under equilibrium conditions the soil slope is acted upon by driving forces and resisting forces, where the driving force is the action of gravity and seepage forces if any, which is resisted by the shear strength of the soil. Soil stability can be defined as the balance of Shear Stress and Shear Strength of the soil, beyond which it fails. Soil

S. Wang et al. (eds.), *New Solutions for Challenges in Applications of New Materials and Geotechnical Issues*, Sustainable Civil Infrastructures,
https://doi.org/10.1007/978-3-319-95744-9_13

failure is defined as the movement of the soil, which is classified into (i) falls (ii) slides and (iii) flows. This phenomenon is influenced by rainfall (Kim et al. 2004; Rahardjo et al. 2005), earthquake (Bray and Travasarou 2007; Hack et al. 2007), pore water pressure (Rinaldi et al. 2004; Smethurst et al. 2006), seepage (Ng and Shi 1998) and surcharge loading (Zhu et al. 2014). Main causes of slope failures are rainfall (Rahimi et al. 2011), Earthquakes (Wilson and Keefer 1984), Seepage (Gasmo et al. 2000), Surcharge load (Sazzad and Haque 2014) and Erosion (Vandamme and Zou 2013). It is vital to understand the characteristics and behavior of soil in a sloped region. Analysis on soil slopes are therefore done to:

- Understand the formation and development of different failures of soil slope and the processes which are responsible for it (Simon et al. 1990),
- Study and monitor the stability of slopes along different scales of time under various conditions,
- Study the soil failure mechanism, type and the environmental factors responsible for the failure,
- Predict the possibility of slope failure for both natural and engineered slopes (Chen and Lee 2003; Saygili and Rathje 2008).

Various analyses are repeatedly performed in order to understand the functional behavior of soil slopes. The prevailing site conditions and the potential mode of failure are considered to be the main factors to be analyzed. The objective of this study is to perform a slope stability analysis on a laboratory scale soil. Natural slopes of varying geometries are scaled down and modeled in a tank of dimensions 2.30 × 1.00 1.25 m. Seepage pressure is induced in the soil mass by promoting a constant inflow rate from an adjacent chamber (Ramkrishnan et al. 2017). The Critical Phreatic Level (the top flow line of a soil mass below which seepage takes place at which the soil fails) for different slope geometries and soil types are found. The experimental results, mainly the CPL for all slope geometries, were analytically modeled using PLAXIS to validate the displacement values obtained from the experimental analysis.

2 Literature Review

Slope failure occurs when the downward movement of soil particles due to gravity and shear stresses exceeds the ultimate shear strength of the soil. The factors that tend to increase the shear stresses or decrease the shear strength increase the chances of failure of a slope. The predominant causes for slope failure are gravity driven groundwater flow, soil topography, effective stresses (Iverson and Reid 1992) etc. Soil topography influences the gravity-driven groundwater flow and the consequent distribution of effective stress, which in turn influences the potential for slope failure. Stability of the slope also depends up on the permeability of the soil as found by Pradel and Raad (1993). Results have shown that reduction in permeability increases the probability of slope saturation, creating the potential for slope failure to occur. Seepage erosion causes liquefaction and rapid slope failures (Crosta and Prisco 1999). The shear strength of the soil can be influenced by transient seepage as stated by Ng and Shi (1998). They concluded that increasing ground water levels result in the decrease of shear strength. It has been observed that groundwater levels rise due to rainfall and antecedent rainfalls have significant influence on the stability of the slope. The effect of transient seepage in an unsaturated soil slope increases the moisture content and soil permeability leading to slope failure (Lam et al. 1987). Soil-water characteristics are strongly dependent on the confining stress which also plays a major role in slope failure (Ng and Pang 2000).

The relative density of the soil sample depends on drop height, particle size of the soil, impact energy, and the impact velocity (Vaid and Negussey 1984). Pluviation can either be air pluviation or water pluviation (Vaid 1984). The relative density achieved through air pluviation depends on the drop height, uniformity of the sand rain, deposition intensity and the average size of the particle (Kolbuszewski 1948; Kolbuszewski and Jones 1961; Butterfield and Andrawes 1970). This method is widely adopted for preparation of large, uniform and repeated soil sample which are not highly cohesive in nature (Dave and Dasaka 2012). Theoretical study by Vaid and Negussey (1988) shows that impact velocity increases non-linearly with height of fall and with increase in particle size. Also, the impact energy increases with increase in height of fall (Kildalen and Stenhamar 1977). Maintaining a constant height of fall and adjusting deposition index while pluviation will be more appropriate for large scale projects (Drnevich et al. 1987; Saussus and Frost 2000; Dupla et al. 2004; Zhao et al. 2006).

The analysis of stability and deformation in geotechnical engineering projects has been made easy after development of a finite element package, PLAXIS 2D. PLAXIS can be used in various areas such as, simulation of foundation excavation and support (Wang et al. 2007), analysis of soil-structure interaction (Viggiani and Tamagnini 2000), slope stability analysis (Hammouri et al. 2008; Alamshahi and Hataf 2009), etc. Evaluation of stability of slopes utilizes stress strain relationships which can be incorporated using PLAXIS as stated by Aryal (2006). An accurate analysis of a slope stability problem and the associated failure mechanism needs to consider the evolving strains and path dependency by means of a constitutive model (Laouafa and Darve 2002). Another approach of soil modeling includes probabilistic modeling of soil profiles which provides quantified information gathered during site investigation and testing, subsurface conditions at a site and provides the basis for predicting performance (Vanmarcke 1977).

3 Materials and Methodology

3.1 Materials

Soil

Three different types of soil found in and around Coimbatore region, Tamil Nadu, India were used for this analysis. Sieve analysis was performed to determine the grain size distribution of the 3 soil types (Fig. 1) and soil sieved through 10 mm sieve (according to IS classification) was used for modeling the slope. Soil 1 was collected from Ettimadai, Coimbatore and Soil 2 and Soil 3 were collected from different sites near

Kanuvai, Coimbatore, Tamil Nadu. The three different types of soil were tested for their basic properties like permeability, shear parameters, particle size, void ratio, specific gravity, maximum dry density, optimum moisture content, etc., and are tabulated in Table 1.

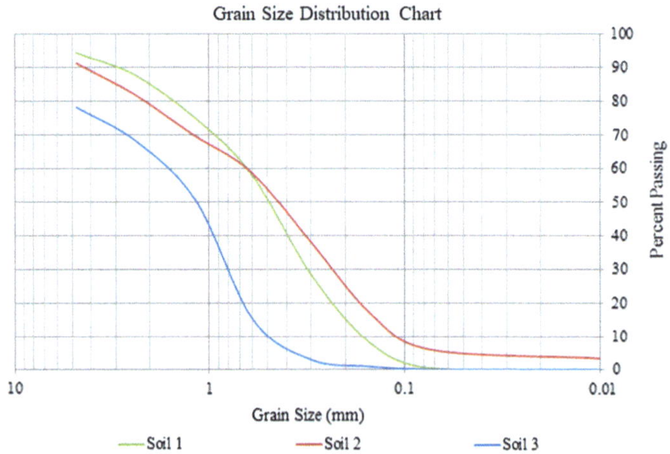

Fig. 1. Grain size distribution for the 3 soils

Table 1. Laboratory experiment results

Parameter	Soil 1	Soil 2	Soil 3
k (m/min)	$8.80e^{-4}$	$3.00e^{-4}$	$4.36e^{-4}$
γ_b (kN/m^3)	23.2	23.6	22.8
G	2.63	2.66	2.42
Φ (degrees)	30.5	41.3	28.9
OMC (%)	11.0	9.00	7.00
C_u	3.22	6.85	3.04
C_c	0.200	0.054	0.151
W_L	14.1	17.5	11.8
W_P	0.00	13.4	18.7
I_P	14.1	6.94	4.05
IS soil classification	SP	SW	SW

3.2 Geotextile

Geotextiles are permeable fabrics made from polypropylene or polyester which, when used in association with soil, can separate, filter, reinforce, protect, or drain. The non-woven geotextile used in this study is shown in Fig. 2 and its properties as provided by the manufacturer are given in Table 2.

Fig. 2. Non-woven geotextile

Table 2. Properties of geotextile

Properties	Standard	200 gsm
Mechanical properties		
Tensile strength	ASTM D 4595	8 kN/m
Elongation	ASTM D 4595	>50%
Physical properties		
Mass/unit area	ASTM D 5261	200 g/m^2
Thickness	ASTM D 5199	1.5 mm
Hydraulic properties		
Flow water head- 5 cm head	ASTM D 4491	50 l/m^2/s
AOS	ASTM D 4751	80 μm

3.3 Tank

A tank with dimensions 2.30 m × 1.00 m × 1.25 m was fabricated using MS metal sheets. One side was provided with windows made of glass, to monitor the slope movement and water level, as shown in Figs. 3 and 4 respectively. The tank was split into two compartments, of length 2.0 m and 0.3 m respectively, using a perforated metal sheet. The perforated sheet was covered with a Non-Woven geotextile material to prevent the movement of sand from one side to the other, but allowing easy movement of water. The larger compartment was used to construct the slope model while the smaller chamber was used as a water tank. This smaller portion was provided with ball valves to regulate the water level in the chamber. The inner surface of the tank was kept smooth to reduce any friction between the tank and the soil, and to ensure its uniform movement within. A graduated scale was fixed on the glass to give an accurate measurement of the water level in the chamber. Grids were drawn inside the tank and on the glass to help in laying the slope accurately and to monitor the slope.

Fig. 3. Tank setup

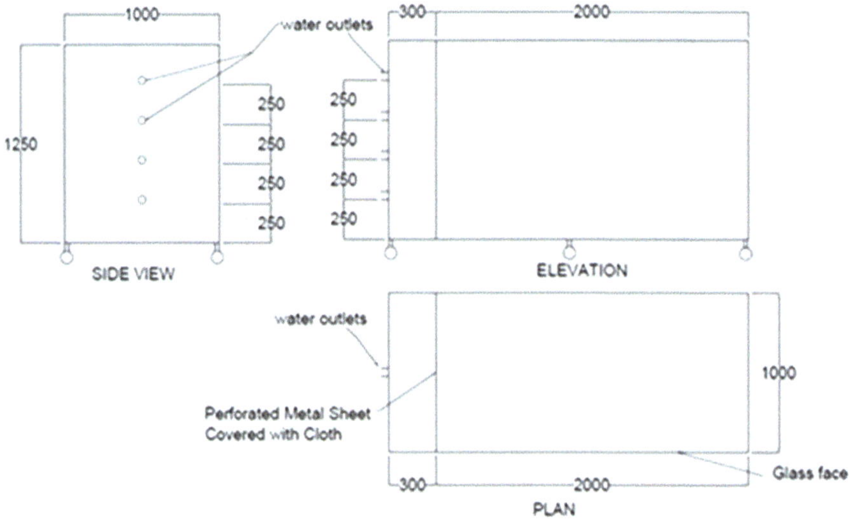

Fig. 4. Tank dimension

3.4 Pluviation Setup

An air pluviation setup, which consists of a pulley system and a pluviation bucket as shown in Fig. 5 was used to model the slopes. The pulley system was designed such that it could be moved throughout the length and breadth of the tank.

3.5 Slope Geometries

A pilot test was conducted to determine the maximum slope angle that could be achieved by air pluviation. The slope was created with a crest height of 0.75 m and a toe height of 0.2 m. The width of the horizontal portion of the crest was altered according to the slope geometry modeled. To get comparable results, the slope angles

Fig. 5. Pluviation setup

chosen were 28°, 30° and 33°. For Soil - 1, the experiment was performed for angles of 28°, 30° and 33°. For Soil - 2 and Soil - 3 the experiments were performed for 28° and 30°. The slope geometries and its dimensions (mm) are shown in Fig. 6 and Table 3 respectively.

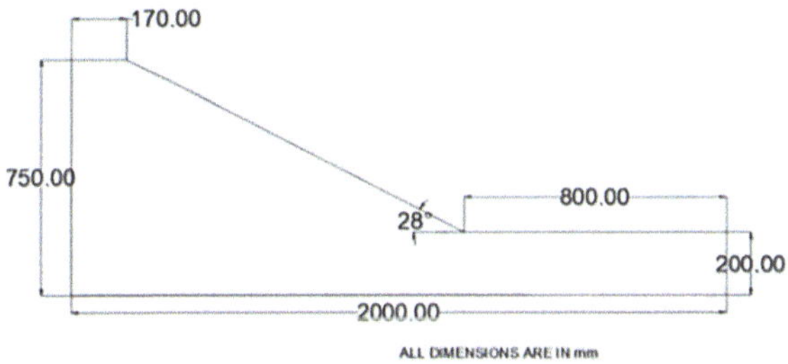

Fig. 6. 28° slope

Table 3. Slope dimensions

Length (mm)			
Slope angle	Crest	Slope	Toe
28°	170	1030	800
30°	250	950	800
33°	353	847	800

3.6 Digital Moisture Sensor (DMS)

The Soil Moisture Sensor, shown in Fig. 7, measures the volumetric water content in the soil. The two probes of the sensor act as a variable resistor – more water in the soil results in better conductivity, lower resistance and higher output voltage. The analog reading varies depending on the voltage used as well as the resolution of the Analogue to Digital Converter (ADC) pins.

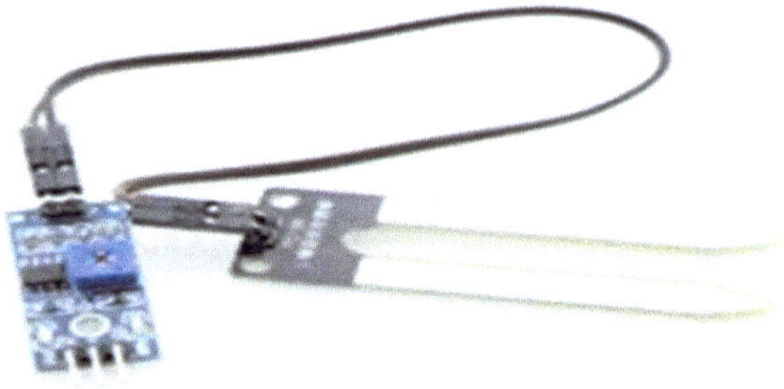

Fig. 7. Digital moisture sensors (DMS)

3.7 Sensor Monitoring Setup

A sensor monitoring setup was installed to monitor the fluctuations in the water level of the soil slope. The position of sensor placement in the soil slope was maintained constant for all slope angles. The sensors were placed at an interval of 20 cm along the length of the tank and 5 cm along the height of the tank. Digital Moisture Sensors as mentioned above were used. The leads were placed in the soil at the abovementioned locations, of which the conductivity increases when water reaches the sensor, making the LED glow in the potentiometer. The indication of LED represents that water has reached the sensor position. A sensor monitoring panel was setup outside the tank to continuously monitor the water flow in real-time, as shown in Fig. 9. A representative sensor placement for the 28° slope geometry is shown in Fig. 8.

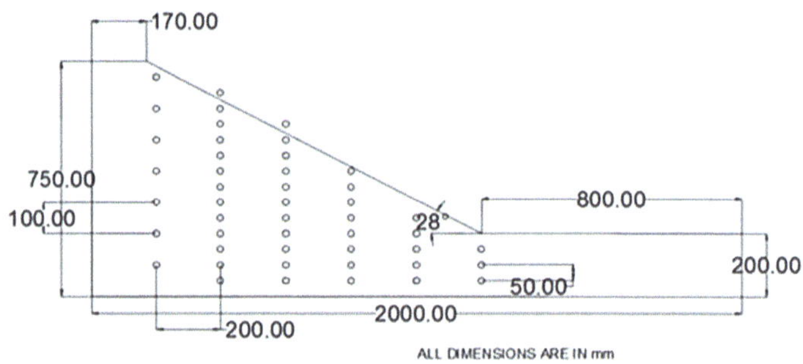

Fig. 8. 28° sensor setup

3.8 Experimental Procedure

The sieved soil sample was poured into the pluviation bucket which was tied to a pulley setup. The height of fall for each soil to achieve the maximum dry density was found by measuring the densities at different fall height. By the to and fro motion of the bucket, soil could flow uniformly through a funnel attached at the bottom. Soil was continuously filled for the first 5 cm and then sensors were placed with an equal interval of 20 cm and 5 cm along length and height of the tank respectively. The process was continued and soil was pluviated to a height of 75 cm at the crest, 20 cm at

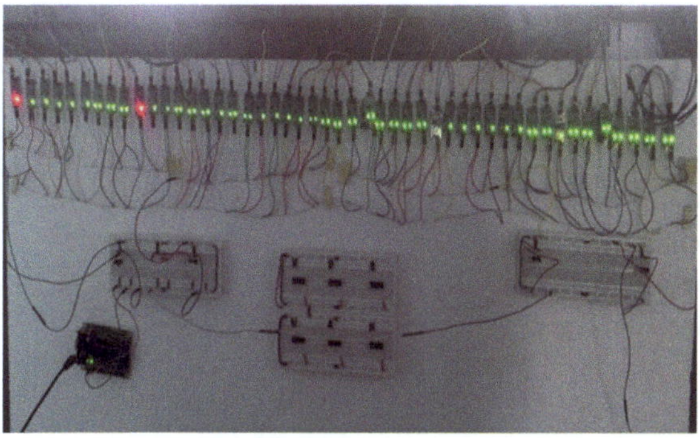

Fig. 9. Sensor monitoring panel

the toe and a length covering 2 m with required slope angle. Dial gauges were fixed at the toe and face of the soil slope to monitor any slope movement. After setting up the model, water with a constant inflow discharge of 2.2 L per minute was induced and maintained throughout the experiment, to simulate a natural groundwater flow condition (Ramkrishnan et al. 2017). Position of the phreatic level and the dial gauge readings were noted for every five minute interval. The experiment continued until the dial gauges showed considerable displacement, indicating ample slope movement or visible slope failure. Outflow discharges were also monitored to ensure steady state flow.

3.9 Numerical Modeling

Numerical studies for the verification of test models were conducted using Finite Element Method (FEM) tools. The plane strain finite element analysis was carried out using PLAXIS 2D software. In this study, the soil was modeled using the Mohr-Coulomb model, which involves five parameters, namely Young's modulus (E), Poisson's ratio (μ), Unit weights, Cohesion(c), and Angle of internal friction (Φ). Finite element analyses were carried out by applying vertical predefined displacements and zero horizontal displacements to the nodes at the top of the soil profile. The specified displacements were applied in equal increments of 250 steps. The soil parameters adopted in all the finite element analyses for the modeled soil slope were tabulated in Table 4. The homogeneous soil slopes of 28° and 30° with induced ground water table were studied extensively and the results are discussed hereafter. The phreatic level at failure - Critical Phreatic Level (CPL), obtained from the experimental analysis was modeled in the software and analyzed. The 28° and 30° slopes for Soil 1 are shown in Figs. 10 and 11 respectively.

Table 4. Parameters of soil

Parameter	Soil 1	Soil 2	Soil 3
Unsaturated unit weight (kN/m^3)	21.1	21.6	21.3
Saturated unit weight (kN/m^3)	22.9	23.3	22.3
Void ratio (e)	0.223	0.205	0.116
Specific gravity (G)	2.63	2.66	2.42
Permeability (k) (m/min)	8.80e^{-4}	3.00e^{-4}	4.36e^{-4}
Stiffness (kN/m^2)	30.0	30.0	30.0
Angle of internal friction (φ)	30.5°	41.3°	28.9°
Unit cohesion (c) (kPa)	24.08	6.26	14.94
Young's modulus (MPa)	40.00	80.00	80.00

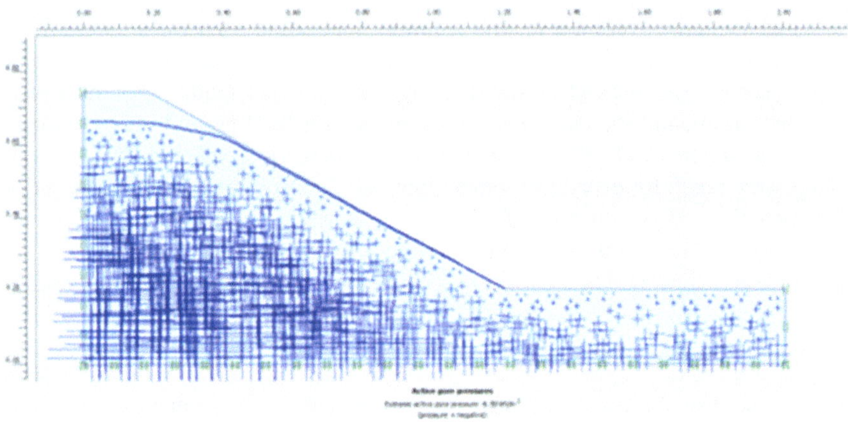

Fig. 10. Critical phreatic level–Soil 1 - 28°

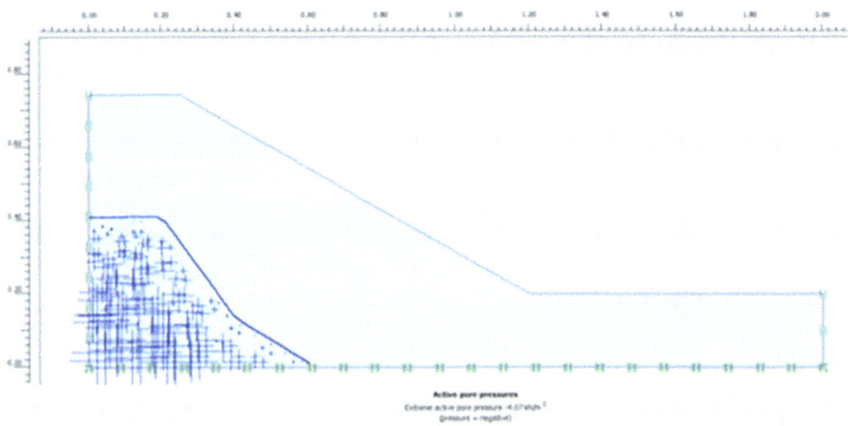

Fig. 11. Critical phreatic level–Soil 2 - 30°

4 Results and Discussion

4.1 Experimental Results

Case 1: Soil 1

Soil - 1 of slope angle 28° with groundwater table

The slope failure occurred after 195 min at a water level of 48 cm from the bottom of the slope. The outflow discharge from the toe region was found to be steady with 1 L per minute after the toe of the slope saturated. The slope had a maximum toe displacement of 0.6 mm. The CPL at failure and slope failure is shown in Figs. 12 and 13.

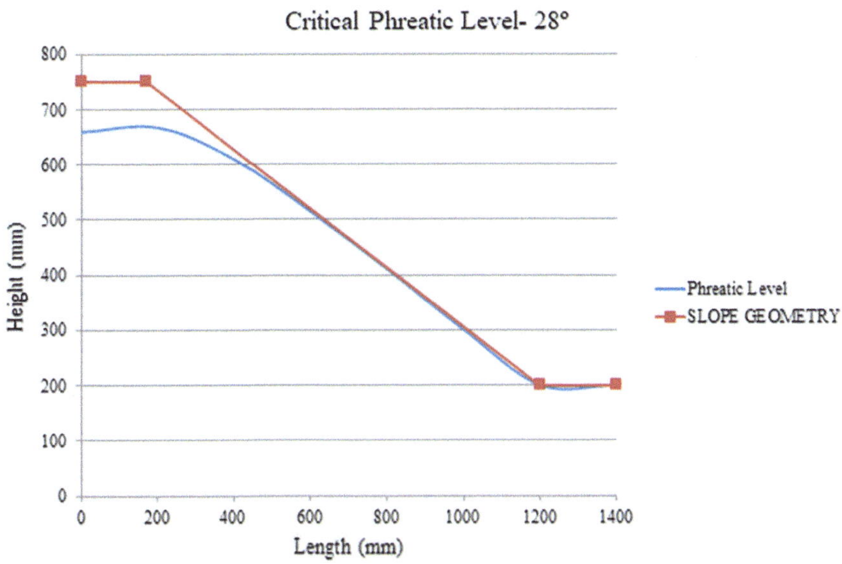

Fig. 12. Critical phreatic level for soil - 1 of slope 28

Fig. 13. Slope failure -28° [soil - 1]

Soil - 1 of slope angle 30° with groundwater table

The slope failure occurred after 160 min at a water level of 46 cm from the bottom of the slope. The outflow discharge from the toe region was found to be steady with 1 L per minute after the toe of the soil slope had saturated. The slope had a maximum toe displacement of 1.02 mm. The CPL at failure and slope failure is shown in Figs. 14 and 15.

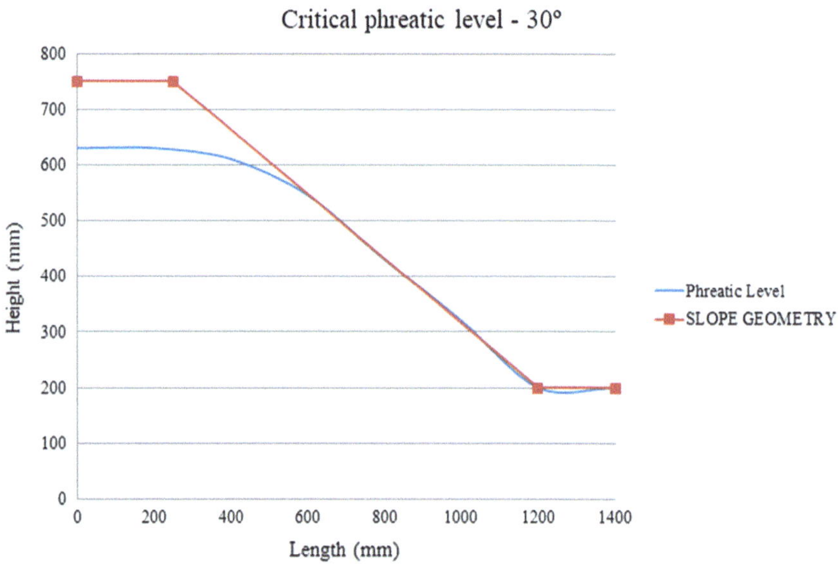

Fig. 14. Critical phreatic level for soil - 1 of slope 30°

Fig. 15. Slope failure - 30° [soil - 1]

Soil - 1 of slope angle 33° with groundwater table
The slope failure occurred after 145 min at a water level of 38.5 cm from the bottom of the slope. The outflow discharge from the toe region was found to be steady with 1.021 L per minute after the toe of the soil slope had saturated. The slope had a maximum toe displacement of 2.15 mm. The CPL at failure and slope failure is shown in Figs. 16 and 17.

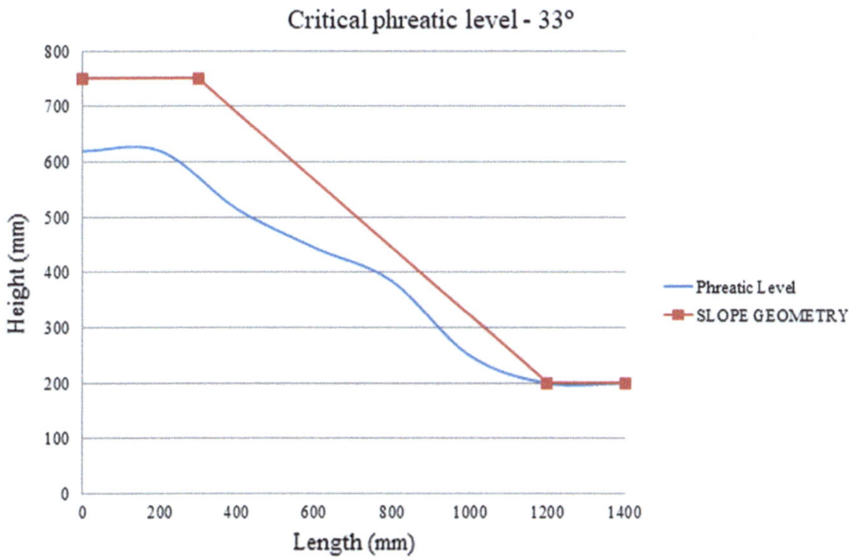

Fig. 16. Critical phreatic level for soil - 1 of slope 33°

Fig. 17. Slope failure - 33° soils - 1

A comparison between different slope angles and their corresponding CPL are shown in Fig. 18 and varying toe displacement is marked against their corresponding slope angles in Fig. 19.

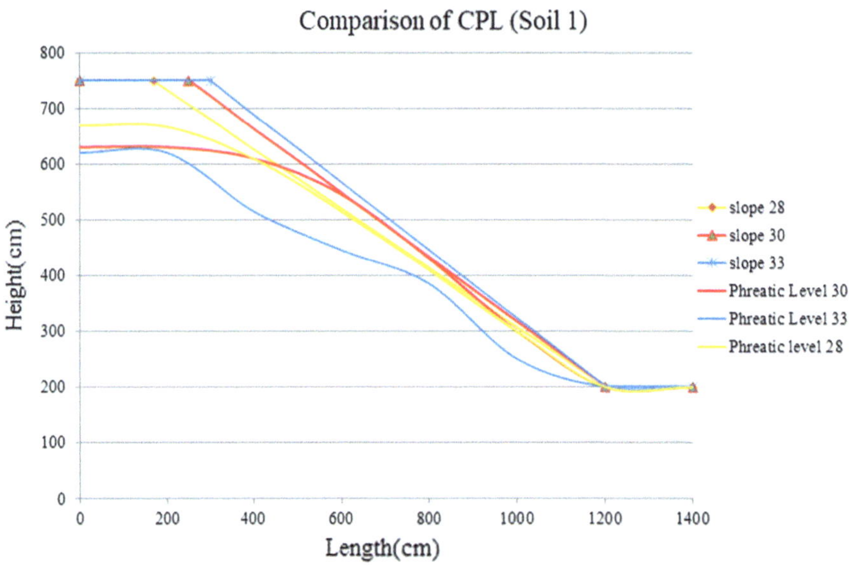

Fig. 18. Different phreatic level for soil - 1

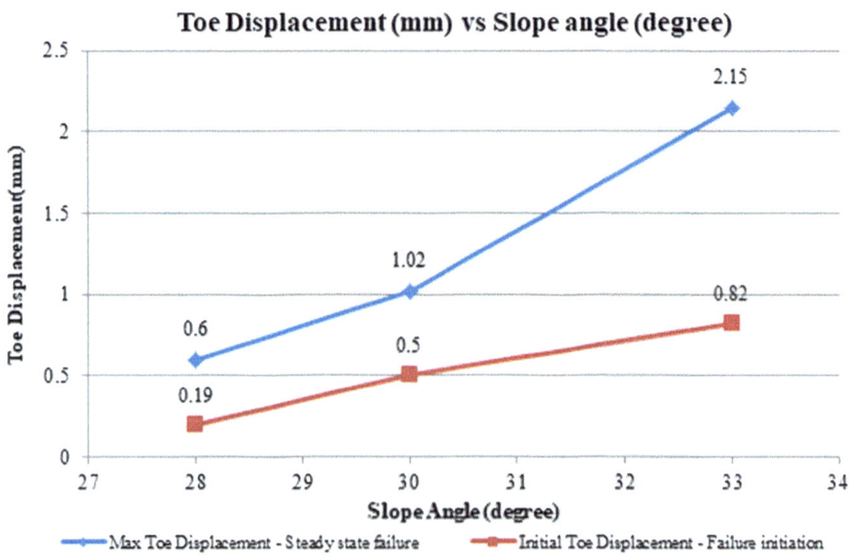

Fig. 19. Toe displacements corresponding to different slope angle

It was observed from the above results that soil with high slope angles fails quicker. CPL for soil with low slope angle was found to be greater than that with steeper slope angles. The soil displacement increased as the slope became steeper.

Case 2: Soil 2

The relatively low permeability and uniform particle size distribution of this soil lead to a considerable capillary rise of water i.e. upward vertical movement of water was faster than the forward horizontal movement. It was observed in this case that the slope failure occurred only when the surface of slope got saturated.

Soil - 2 of slope angle 28° with groundwater table

The failure occurred after 85 min at a water level of 38 cm from the bottom of the slope, recording a face displacement of 4.03 mm. The ultimate failure was observed as soil flow, which occurred after the saturation of the top surface of the slope. The CPL and slope failure is shown in Figs. 20 and 21 respectively.

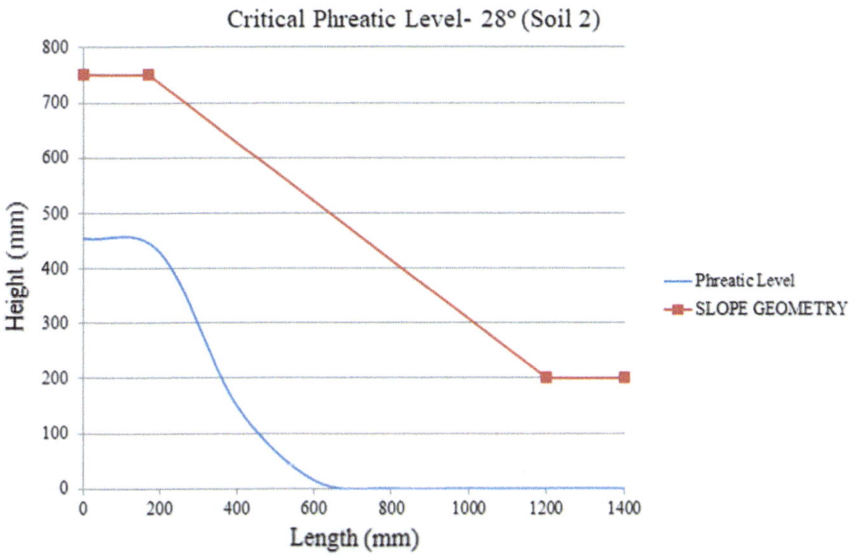

Fig. 20. Critical phreatic level for soil - 2 of slope 28°

Fig. 21. Slope failure - 28° [soil - 2]

Soil - 2 of slope angle 30° with groundwater table

The failure occurred after 75 min at a water level of 38 cm from the bottom of the slope, recording a face displacement of 2.08 mm. The ultimate failure was soil flow which occurred after the saturation of the top surface of the slope and the maximum final face displacement was found to be 5.18 mm. The CPL and slope failure is shown in Figs. 22 and 23 respectively.

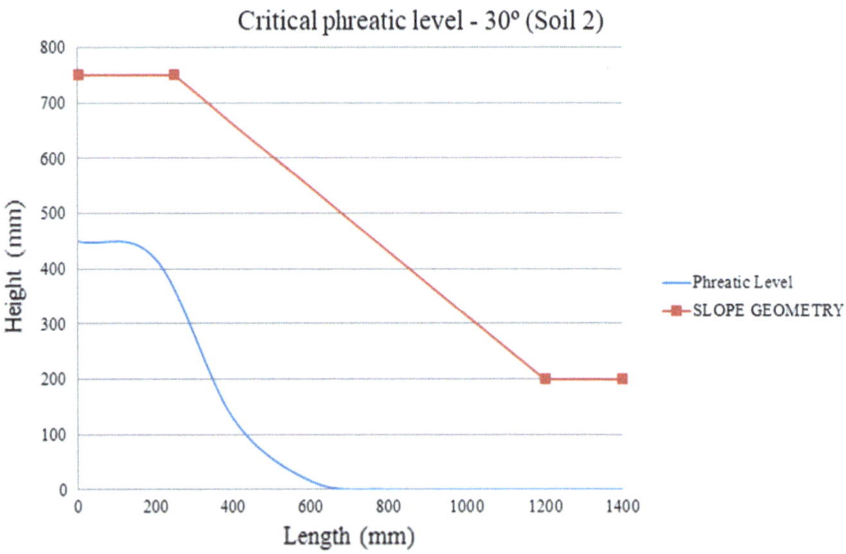

Fig. 22. Critical phreatic level for soil - 2 of slope 30°

Fig. 23. Slope failure - 30° [soil - 2]

As a considerable amount of water moved in the soil slope, there occurred a series of displacements along the face of the soil slope. The ultimate failure occurred when the top of the slope saturated and caused earth flow in both slope geometries.

Case 3: Soil 3

Soil - 3 of slope angle 28° with groundwater table

The failure occurred after 115 min and the water table was found to be at 43.5 cm from the bottom of the slope. The failure of the slope was observed as considerable settlement and face displacement. Displacement at critical failure was recorded to be 22.1 mm of face displacement and maximum final face displacement was found to be 24.31 mm. The CPL and slope failure is shown in Figs. 24 and 25 respectively.

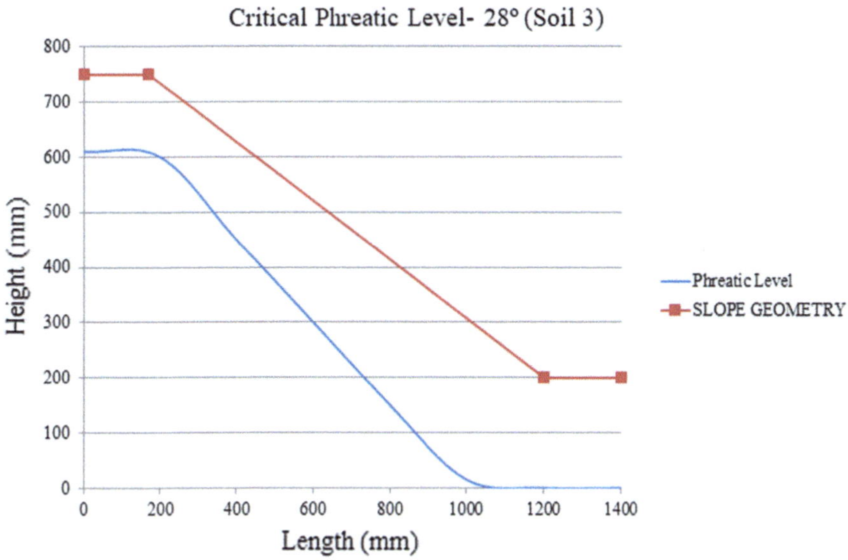

Fig. 24. Critical phreatic level for soil - 3 of slope 28°

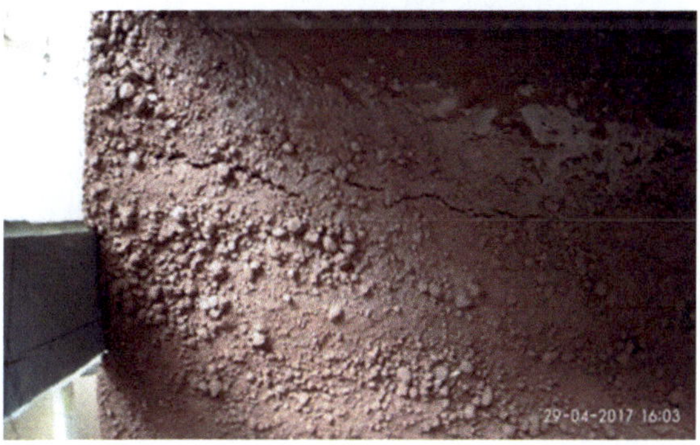

Fig. 25. Slope failure - 28° [soil - 3]

Soil - 3 of slope angle 30° with groundwater table

The failure occurred after 110 min and the water table was found to be 44 cm from the bottom of the slope. The major failure of the slope was due to face displacement. Displacement at critical failure was recorded to be 13.39 mm of face displacement and maximum final face displacement was found to be 17.79 mm. The CPL and slope failure is shown in Figs. 26 and 27 respectively.

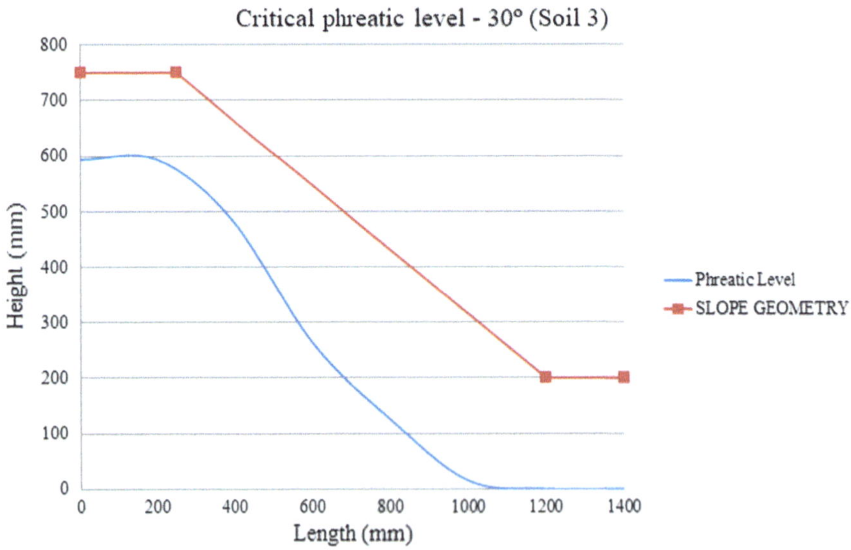

Fig. 26. Critical phreatic level for soil - 3 of slope 30°

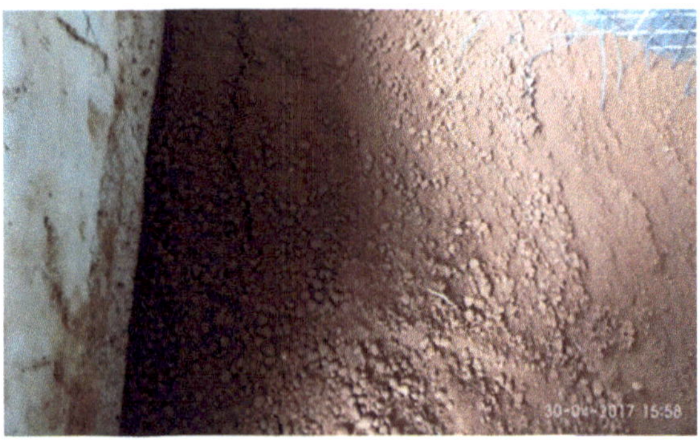

Fig. 27. Slope failure - 30° [soil - 3]

Both the slope geometries of Soil - 3 were subjected to high face displacement as the water flowed through the slope. The ultimate failure of the slopes was marked by large displacements and earth movements under the action of water.

4.2 Comparison of CPLs for Different Angles of the Same Soil

By comparing the CPLs for different angles of the same soil, it was evident that the CPL was lower for steeper slopes. The CPLs for different angles and different soils are shown in Figs. 18, 28 and 29. It was observed that steeper slopes have greater displacement and fail quicker than gradual slopes, provided the different slopes are of same density and have same inflow rate of water. CPL for different soils mainly depend on the permeability and particle size distribution of soil (coefficient of curvature and coefficient of uniformity), which in turn increases the seepage pressures and total driving moments generated, reducing the mobilization of maximum shear strength along the slip surfaces. The nature of slope failure also depends on the particle size distribution of the soil.

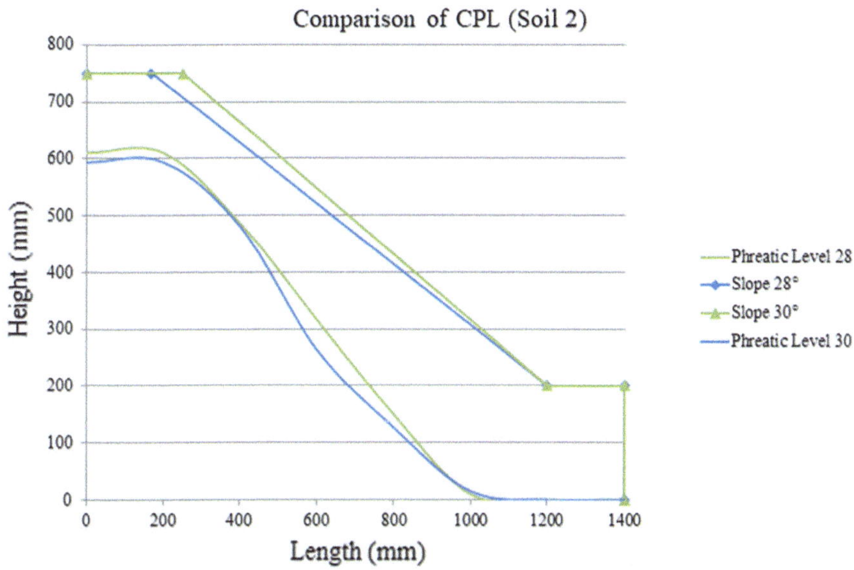

Fig. 28. Different phreatic level for soil - 2

4.3 Comparison of CPLs for Different Soils at Same Slope Angle

The CPLs of different soils varied from each other though they had similar slope geometries. The displacements happened at the toe region or on the slope face, varying from small displacements of 2 mm to much higher displacements of 20 mm. Depending on various characteristics and properties of the soil, the water movement at the specified inflow rate dominated in either vertical upward movement i.e. capillary rise or forward horizontal movement, causing different modes of failure too. The CPLs of different soils of same slope geometries are graphically compared in Figs. 30 and 31.

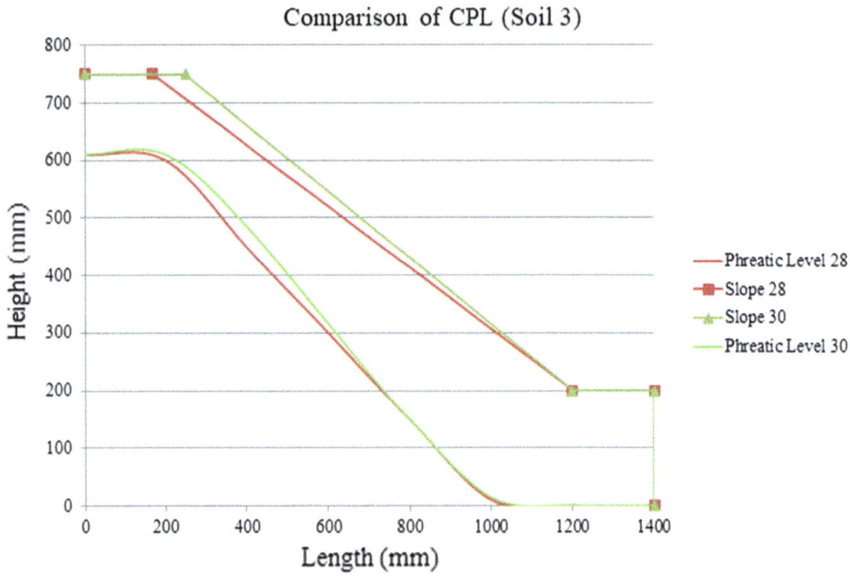

Fig. 29. Different phreatic level for soil - 3

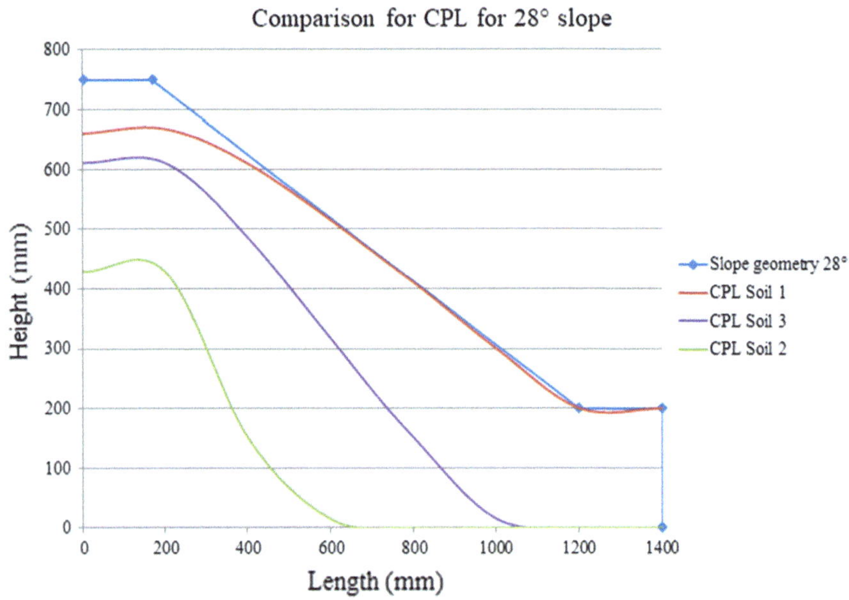

Fig. 30. Critical phreatic levels for slope angle 28°

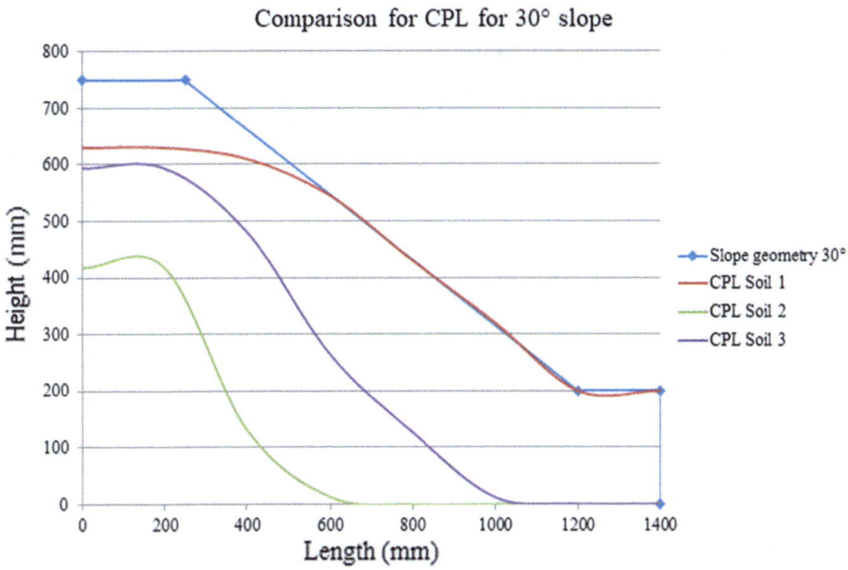

Fig. 31. Critical phreatic levels for slope angle 30°

5 Numerical Results

The experimental results were modeled in PLAXIS and validated. It was found that the failure patterns and displacements observed in the software and the experimental studies were similar. A possible explanation for the minor difference between the experimental and numerical displacement results could be the fact that PLAXIS 2D runs a plastic analysis for the slope geometries until it reaches ultimate failure, incorporating automatic time steps. However, in the experimental studies, the displacement was noted at the time of failure and the tests were run only for a limited duration. The displacement values reported from the numerical analysis are the maximum values attained. However, if the displacements are considered at the location where the displacement was measured in the experimental study, the difference between the two values would be much lesser. The failure patterns for different slope geometries for different types of soils are discussed below.

5.1 Homogenous Soil Slope with Ground Water Table

After the CPLs from the experimental results were simulated in the software, the failure displacement diagrams for 28° and 30° slopes were obtained and are represented in Table 5 and Fig. 32. Slight differences in the total displacements of experimental and numerical models may be due to the difference in the field density achieved in experimental and numerical study. The total displacement diagrams from FEM analysis for different soils at 30° slope angles are shown in Figs. 33, 34 and 35. The failure diagrams and displacement values obtained from FEM analyses for Cases 1 and 2 are similar to the failure patterns observed in the experimental studies, thereby validating the experimental procedure and the sensor readings for the phreatic levels. From the experimental analysis and the subsequent analytical validation, it was observed that the CPL and the slope failure parameters depended on the soil properties such as soil group classification, coefficient of curvature and uniformity, etc. This analysis can thus be carried forward to obtain a reliable method that uses the aforementioned soil parameters to predict the CPL and slope failure parameters such as the mode of failure, maximum displacement, etc.

Table 5. Displacement (mm) – experimental and numerical results

Total displacement (mm)		
Slope angle	Experimental results	Numerical results
Soil 1		
28°	0.19	0.10
30°	0.50	0.20
Soil 2		
28°	3.69	30.47
30°	2.08	32.31
Soil 3		
28°	22.10	40.59
30°	13.39	40.74

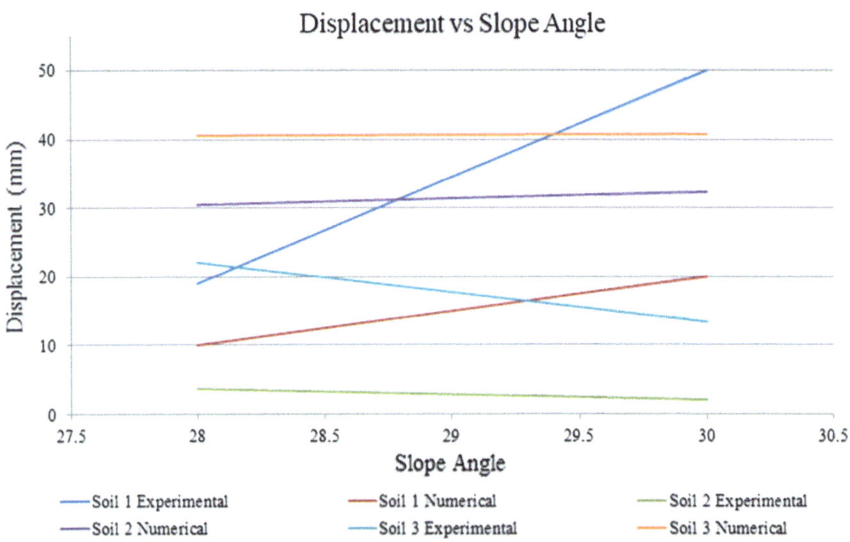

Fig. 32. Displacement vs slope angle for different soil types

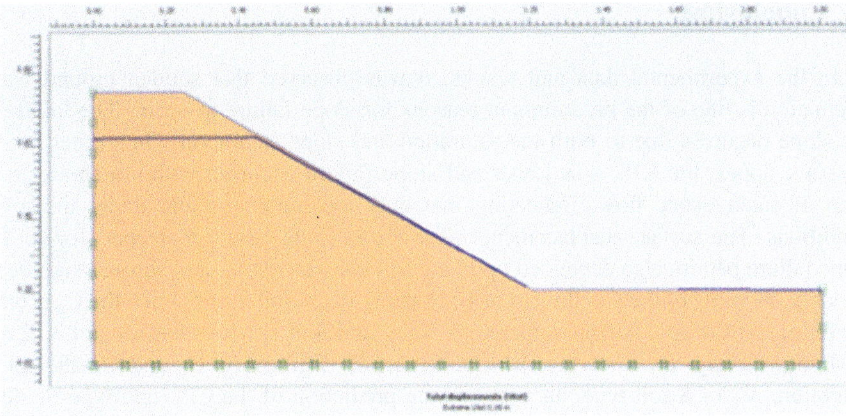

Fig. 33. Soil - 1 total displacement

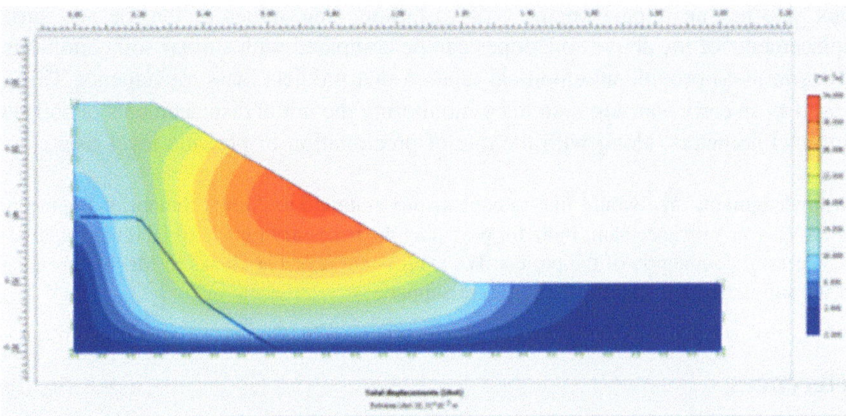

Fig. 34. Soil - 2 total displacements

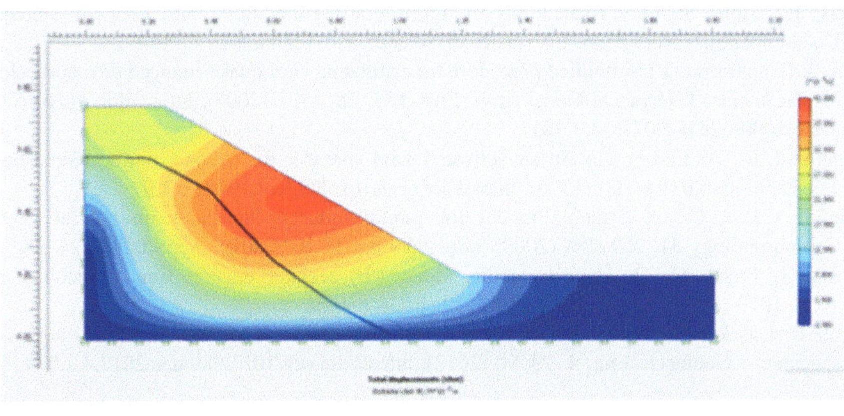

Fig. 35. Soil - 3 total displacements

6 Conclusion

From the experimental data and results, it was observed that sudden ground water fluctuation is one of the predominant reasons for slope failure to occur. The failure of the slope occurred due to both toe saturation and slope saturation. On steeper homogeneous slopes, the CPL was lower and slope failure occurred within a shorter time span of steady state flow, indicating that they are more unstable under the given conditions. The surface displacements were found to be larger in steeper slopes. The slope failure pattern also depended upon the soil characteristics, and showed significant variation in terms of the Coefficient of Uniformity (C_u). Soil 2, for which the C_u is 6.85, the failure was drastic when compared to Soil 1 and Soil 3, which has a C_u of 3.22 and 3.04 respectively. By knowing the Coefficient of Uniformity, C_u and Coefficient of Curvature, C_c of a soil type, an approximate prediction of the CPL and type of slope failure can be done. The failure occurred upon toe saturation in the case of soil slopes with highly plastic soil, whereas for slopes having soil with low plasticity index, the failure took place before the water reached the toe. We can also conclude that plasticity index also has an impact on the type of failure. The values of the toe and surface displacements of the above soil slopes can be compared with similar soil conditions in hilly terrains to provide an empirical relation that predicts mass movements. This can be used as an early warning system by monitoring the initial displacements of soil slope using inclinometers, along with the rate of precipitation or phreatic level rise.

Acknowledgment. We would like to express our gratitude to Amrita School of Engineering, Amrita Vishwa Vidyapeetham, India for providing the necessary funds and facilities required for the successful completion of the project. We express our heartfelt thanks to Mr. K. Sreenivasan and the supporting lab staff for their sincere support.

References

Alamshahi, S., Hataf, N.: Bearing capacity of strip footings on sand slopes reinforced with geogrid and grid-anchor. Geotext. Geomembr. **27**, 217–226 (2009). https://doi.org/10.1016/j.geotexmem.2008.11.011

Aryal, K.: Slope Stability Evaluations by Limit Equilibrium and Finite Element Methods. Doctoral Thesis, Norwegian University of Science and Technology (2006)

Bray, J., Travasarou, T.: Simplified procedure for estimating earthquake-induced deviatoric slope displacements. J. Geotech. Geoenviron. Eng. **133**, 381–392 (2007). https://doi.org/10.1061/(asce)1090-0241(2007)133:4(381)

Butterfield, R., Andrawes, K.: An air activated sand spreader for forming uniform sand beds. Géotechnique **20**, 97–100 (1970). https://doi.org/10.1680/geot.1970.20.1.97

Chen, H., Lee, C.: A dynamic model for rainfall-induced landslides on natural slopes. Geomorphology **51**, 269–288 (2003). https://doi.org/10.1016/s0169-555x(02)00224-6

Crosta, G., Prisco, C.: On slope instability induced by seepage erosion. Can. Geotech. J. **36**, 1056–1073 (1999). https://doi.org/10.1139/cgj-36-6-1056

Dave, T., Dasaka, S.: Assessment of portable traveling pluviator to prepare reconstituted sand specimens. Geomech. Eng. **4**, 79–90 (2012). https://doi.org/10.12989/gae.2012.4.2.079

Drnevich, V., Rad, N., Tumay, M.: Factors affecting sand specimen preparation by raining. Geotech. Test. J. **10**, 31–37 (1987). https://doi.org/10.1520/gtj10136j

Dupla, J., Canou, J., Gouvenot, D.: An advanced experimental set-up for studying a monodirectional grout injection process. Ground Improv. **8**, 91–99 (2004). https://doi.org/10.1680/grim.8.3.91.41117

Gasmo, J., Rahardjo, H., Leong, E.: Infiltration effects on stability of a residual soil slope. Comput. Geotech. **26**, 145–165 (2000). https://doi.org/10.1016/s0266-352x(99)00035-x

Hack, R., Alkema, D., Kruse, G., et al.: Influence of earthquakes on the stability of slopes. Eng. Geol. **91**, 4–15 (2007). https://doi.org/10.1016/j.enggeo.2006.12.016

Hammouri, N., Malkawi, A., Yamin, M.: Stability analysis of slopes using the finite element method and limiting equilibrium approach. Bull. Eng. Geol. Environ. **67**, 471–478 (2008). https://doi.org/10.1007/s10064-008-0156-z

Iverson, R., Reid, M.: Gravity-driven groundwater flow and slope failure potential: 1. Elastic effective-stress model. Water Resour. Res. **28**, 925–938 (1992). https://doi.org/10.1029/91wr02694

Kildalen, S., Stenhamar P.: NGI laboratory sand rainer. Internal report 51505–15, Norwegian Geotechnical Institute (1977)

Kim, J., Jeong, S., Park, S., Sharma, J.: Influence of rainfall-induced wetting on the stability of slopes in weathered soils. Eng. Geol. **75**, 251–262 (2004). https://doi.org/10.1016/j.enggeo.2004.06.017

Kolbuszewski, J.: An experimental study of the maximum and minimum porosities of sands. In: Proceedings of the Second International Conference of Soil Mechanics and Foundation Engineering, vol. 1, pp. 158–165 (1948)

Kolbuszewski, J., Jones, R.: The preparation of sand samples for laboratory testing. Proc. Midl. Soil Mech. Found. Eng. Soc. **4**, 107–123 (1961)

Lam, L., Fredlund, D., Barbour, S.: Transient seepage model for saturated–unsaturated soil systems: a geotechnical engineering approach. Can. Geotech. J. **24**, 565–580 (1987). https://doi.org/10.1139/t87-071

Laouafa, F., Darve, F.: Modelling of slope failure by a material instability mechanism. Comput. Geotech. **29**, 301–325 (2002). https://doi.org/10.1016/s0266-352x(01)00030-1

Ng, C., Pang, Y.: Influence of stress state on soil-water characteristics and slope stability. J. Geotech. Geoenviron. Eng. **126**, 157–166 (2000). https://doi.org/10.1061/(asce)1090-0241(2000)126:2(157)

Ng, C., Shi, Q.: A numerical investigation of the stability of unsaturated soil slopes subjected to transient seepage. Comput. Geotech. **22**, 1–28 (1998). https://doi.org/10.1016/s0266-352x(97)00036-0

Pradel, D., Raad, G.: Effect of permeability on surficial stability of homogeneous slopes. J. Geotech. Eng. **119**, 315–332 (1993). https://doi.org/10.1061/(asce)0733-9410(1993)119:2(315)

Rahardjo, H., Lee, T., Leong, E., Rezaur, R.: Response of a residual soil slope to rainfall. Can. Geotech. J. **42**, 340–351 (2005). https://doi.org/10.1139/t04-101

Rahimi, A., Rahardjo, H., Leong, E.: Effect of antecedent rainfall patterns on rainfall-induced slope failure. J. Geotech. Geoenviron. Eng. **137**, 483–491 (2011). https://doi.org/10.1061/(asce)gt.1943-5606.0000451

Ramkrishnan, R., Karthik, V., Unnithan, M.S., Kiran Balaji, R., Athul Vinu, M., Venugopalan, A.: Stabilization of seepage induced soil mass movements using sand drains. Geotech. Eng. J. SEAGS & AGSSEA, **48**(4), 129–137 (2017)

Rinaldi, M., Casagli, N., Dapporto, S., Gargini, A.: Monitoring and modelling of pore water pressure changes and riverbank stability during flow events. Earth Surf. Proc. Land. **29**, 237–254 (2004). https://doi.org/10.1002/esp.1042

Saussus, D., Frost, J.: Simulating the membrane contact patterns of triaxial sand specimens. Int. J. Numer. Anal. Meth. Geomech. **24**, 931–946 (2000). https://doi.org/10.1002/1096-9853(200010)24:12<931:aid-nag100>3.0.co;2-4

Saygili, G., Rathje, E.: Empirical predictive models for earthquake-induced sliding displacements of slopes. J. Geotech. Geoenviron. Eng. **134**, 790–803 (2008). https://doi.org/10.1061/(asce)1090-0241(2008)134:6(790)

Sazzad, M., Haque, M.: Effect of surcharge on the stability of slope in a homogeneous soil by FEM. In: 2nd International Conference on Advances in Civil Engineering, pp 315–318 (2014)

Simon, A., Larsen, M., Hupp, C.: The role of soil processes in determining mechanisms of slope failure and hillslope development in a humid-tropical forest eastern Puerto Rico. Geomorphology **3**, 263–286 (1990). https://doi.org/10.1016/0169-555x(90)90007-d

Smethurst, J., Clarke, D., Powrie, W.: Seasonal changes in pore water pressure in a grass-covered cut slope in London Clay. Géotechnique **56**, 523–537 (2006). https://doi.org/10.1680/geot.2006.56.8.523

Vaid, Y.: Relative density of pluviated sand samples. Jpn. Soc. Soil Mech. Found. Eng. **24**(2), 101–105 (1984)

Vaid, Y., Negussey, D.: Relative density of pluviated sand samples. Soils Found. **24**, 101–105 (1984). https://doi.org/10.3208/sandf1972.24.2_101

Vaid, Y., Negussey, D.: Preparation of reconstituted sand specimens. Advanced triaxial testing of soils and rock, ASTM STP977. In: Donaghe, R.T., Chaney, R.C., Silver, M.L., (eds) ASTM International, West Conshohocken, PA, pp. 405–417 (1988)

Vandamme, J., Zou, Q.: Investigation of slope instability induced by seepage and erosion by a particle method. Comput. Geotech. **48**, 9–20 (2013). https://doi.org/10.1016/j.compgeo.2012.09.009

Vanmarcke, E.: Probabilistic modeling of soil profiles. J. Geotech. Eng. Div. **103**(11), 1227–1246 (1977)

Viggiani, G., Tamagnini, C.: Ground movements around excavations in granular soils: a few remarks on the influence of the constitutive assumptions on FE predictions. Mech. Cohesive-frictional Mater. **5**, 399–423 (2000). https://doi.org/10.1002/1099-1484(200007)5:5<399:aid-cfm101>3.0.co;2-r

Wang, J., Wang, C., Lu, G.: Application of PLAXIS to simulation of foundation excavation and support. Chin. J. Rock Mech. Eng. **35** (2007)

Wilson, R., Keefer, D.: Dynamic analysis of a slope failure from the 6 August 1979 Coyote Lake, California, earthquake. Int. J. Rock Mech. Min. Sci. Geomech. Abstr. **21**, 220–221 (1984). https://doi.org/10.1016/0148-9062(84)90499-6

Zhao, Y., Gafar, K., Elshafie, M., Deeks, A., Knappett, J., Madabushi, S.: Calibration and use of new automatic sand pourer. In: Sixth International Conference on Physical Modeling in Geotechnics, Hong Kong, 4–6 August, Taylor & Francis, London, pp. 265–270 (2006)

Zhu, H., Shi, B., Zhang, J., et al.: Distributed fiber optic monitoring and stability analysis of a model slope under surcharge loading. J. Mt. Sci. **11**, 979–989 (2014). https://doi.org/10.1007/s11629-013-2816-0

Study on Surface Deformation Caused by Comprehensive Mining in a Coal Mine in Northwest China

Jie Liu[1], Zheng Lu[1(✉)], Hailin Yao[1], and Liming Lang[2]

[1] State Key Laboratory of Geomechanics and Geotechnical Engineering, Institute of Rock and Soil Mechanics, Chinese Academy of Sciences, Wuhan 430071, China
{liujie_whrsm, lzwhrsm}@163.com, hlyao@whrsm.ac.cn

[2] Sanjiang Engineering Construction Administration Bureau of Heilongjiang Province, Haerbin 150081, China
18945084966@163.com

Abstract. The exploitation of underground coal can cause many problems such as the ground surface subsidence, ground crack, surface damage of the road, and it would affect people travelling, living and life safety seriously. Based on this situation, this paper will point at a typical coal exploitation (including a large-scale coal exploitation and private coal mine exploitation), study the change mechanisms of the surface deformation and the law of the rock movement of underground coal mining conditions, surface stability and safety are verified by the ground surface settlement and horizontal displacement, The result shows that: The mining subsidence caused by the large coal mining has some influences to the highway operation, the surface deformation is stacked by the exploitation of two coal mining, the damage of road surface is mainly caused by the private coal mine exploitation, the long term deformation caused by the large coal mine is not the main influence of the damage of the road and village. The influence of the coal mining is small and the deformation is small, the building does not need maintenance, and can be used normally.

Keywords: Coal mine · Vertical subsidence · Horizontal displacement
Mining surface · Rock movement

1 Introduction

The national infrastructure highlights the problem of the scarcity of land resources. Therefore, many large factories, railway and highway construction should be built above the goaf. It is very probable that the overlying rock of goaf are broken and collapsed under the external loads, which would affect the stability of foundation and lead to instability and failure of the building. The squeeze pressure caused by the mining is transmitted to the road surface through the subgrade, which makes the moving vehicle running in the air, causing the rollover accident. And some underground mining may produce the additional stress, resulting in the house deformation exceed its resistance ability, the house is damaged. So it is important to study the

S. Wang et al. (eds.), *New Solutions for Challenges in Applications of New Materials and Geotechnical Issues*, Sustainable Civil Infrastructures,
https://doi.org/10.1007/978-3-319-95744-9_14

surface deformation caused by comprehensive mining in a coal mine to prevent the occurrence of disasters. Many scholars have studied the problem and got some achievements, some of them studied the deformation of the ground in the process of mining, some of them studied the long term ground permanent deformation after mining, their studies show that the ground deformation can be very small when the support structure of the mine has enough strength, and it also has no influences to the ground building and road. This project is special: a large mine is deep in the ground, a small private mining is in a shallow position in the ground, and the private mining is short of enough supporting measures, so it would have some adverse conditions of the ground deformation, it is our research emphases.

2 Stability Analysis of Mining Area

2.1 Engineering Problems

In this article a China coal mine which locates in the northwest are studied. There is a private coal mining in the upper part, the depth of the large coal mining is more than 500 m, the exploitation of underground coal mining and exploitation of the private coal mines, resulting in local surface subsidence, building and road crack which and affect the mining area residents living and travelling safety seriously, it can be seen in Fig. 1.

Fig. 1. Site condition of the ground above the coal mine

This paper will carry out numerical calculation to study the influence of underground coal mining to the ground surface.

2.2 Selection of Physical and Mechanical Parameters of Stratum

The average mining length of the coal mining face is 830 m, the dip angle of coal seam is 15°– 48°, the length of the coal seam which perpendicular to the strike length is about 35 m. the calculate boundary range is 1630 m along the direction of coal seam and is 950 m along the vertical direction. In this paper, two sections are set up respectively along the strike direction and vertical direction, and finite element calculation are done to study the influence of underground coal mining to the ground surface.

The parameters of rock and soil layer in the study area is in Table 1 (The parameters come from indoor physical mechanical tests and some related literature), a thrust fault is in the study area, its dip angle is 30° – 60°, vertical drop is 42 m and the horizon drop is 2 m, in the calculation, the thrust fault is considered.

2.3 Model 1

2.3.1 Calculation Range and Boundary Conditions

The selected section is vertical to the working surface and locates in the center of the exploitation basins, the calculation result based on this section is reasonable to evaluate the surface safety. The calculation range, boundary conditions and finite element mesh of model 1 are shown in Fig. 2. The element is divided into 4689 parts and 5150 units.

2.3.2 Calculation Condition

Case 1: Half of the coal seam of the upper private coal mine has been exploited.
Case 2: Large coal mining face has been exploited completely.

2.3.3 Calculation Process

The principle of simulation calculation is to keep the calculation steps consistent with the actual construction process, and make sure the results reliable. The calculation procedure is as follows:

(1) Using the dead weight of rock and soil to simulate initial stress.
(2) Simulating the process of coal seam mining, the top coal collapse with coal mining, In the simulation, each calculate step of the coal seam mining is 1.2 m.
(3) Simulating roof falling process. The roof falling with the work face moving forward 7 m.

2.3.4 Calculation Result

In accordance to the calculation process, there are 84 nodes on the ground surface from A to B, and the ground surface horizontal displacement and vertical settlement situation are shown in Figs. 3 and 4.

The analysis of horizontal displacement and vertical settlement curve are shown in Figs. 3 and 4. The curves of horizontal displacement and vertical settlement are shown in Figs. 5, 6, 7 and 8.

In case 1, the small coal mining has been exploited, the center of the surface subsidence basin locates above the private coal mine, the maximum horizontal displacement in the northwest edge of the basin is 0.85 cm, the maximum horizontal displacement in the southeast edge of basin is 0.84 cm. The maximum vertical settlement of the surface is 1.98 cm.

In case 2, the large coal mine has been exploited completely, the maximum horizontal displacement in the northwest edge of the surface basin to the southeast is 4.39 cm, and the maximum horizontal displacement in the northeast of the surface basin is 3.30 cm. and the maximum settlement in the center of the subsidence basin is 11.10 cm.

Table 1. Physical and mechanical parameters of the model

Layer	Rock layer	Thickness (m)	Density ($kg \cdot m^{-3}$)	Elastic modulus (MPa)	Poisson ratio (μ)	Cohesion (MPa)	Internal friction angle (°)
1	Loess	10.5	1700	55	0.3	0.03	20
2	Sandy mudstone	73	2300	8000	0.255	3.5	33
3	Sandstone interbedded with mudstone	33.5	2300	13,060	0.25	21	44
4	Coal seam	3	1390	2000	0.3	2.8	28
5	Sandstone interbedded with mudstone	29.9	2300	13,060	0.25	21	44
6	Siltstone	41.1	2250	13,000	0.23	18	40
7	Sandstone interbedded with siltstone	39	2300	13,060	0.25	21	44
8	Sandy mudstone	26.5	2300	8000	0.255	3.5	33
9	Medium grained sandstone	29.6	2400	12,000	0.22	21	44
10	Sandstone interbedded with medium grained sandstone	35.12	2300	13,060	0.25	21	44
11	Sandstone	231.72	2300	13,060	0.25	21	44
12	Grit	4.9	2600	12,000	0.23	19	37
13	Siltstone	3.6	2250	13,000	0.23	18	40
14	Coal seam	7.7	1390	2000	0.3	2.8	28
15	Argillaceous siltstone	2.5	2300	13,000	0.23	18	40
16	Coarse sandstone	5	2600	12,000	0.23	19	37
17	Coal	1.8	1390	2000	0.3	2.8	28
18	Siltstone	16.1	2250	13,000	0.23	18	40
19	Coal seam	4.8	1390	2000	0.3	2.8	28
20	Coarse sandstone	50	2600	12,000	0.23	19	37

After the completion of mining face, comparing case 1 with case 2 it can be seen that: the subsidence basin range and deformation increases significantly, the maximum settlement of ground surface deformation increases about 9.12 cm.

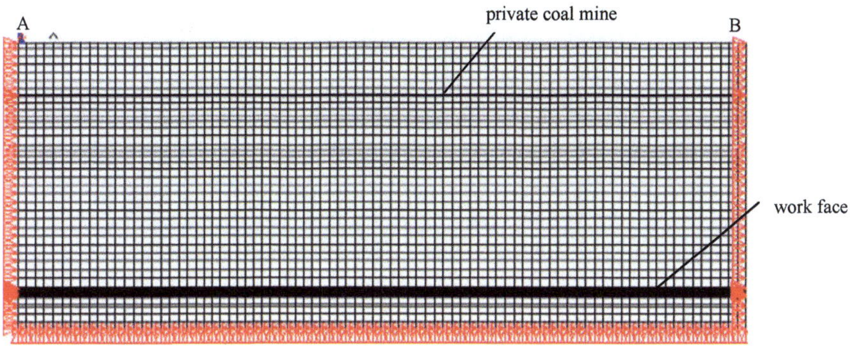

Fig. 2. The boundary condition and element partition of model 1

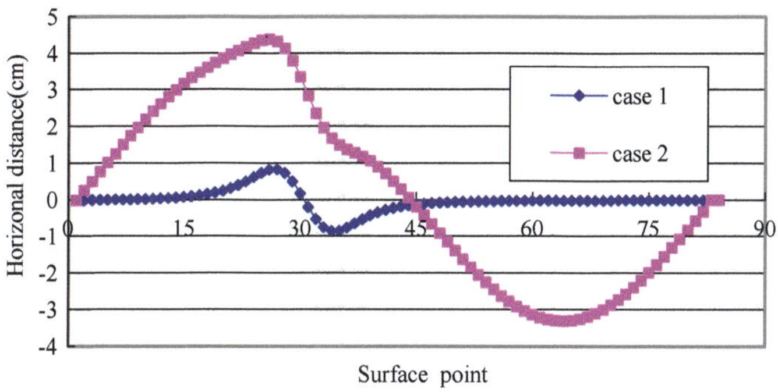

Fig. 3. The surface horizon displacement change curves of two work conditions

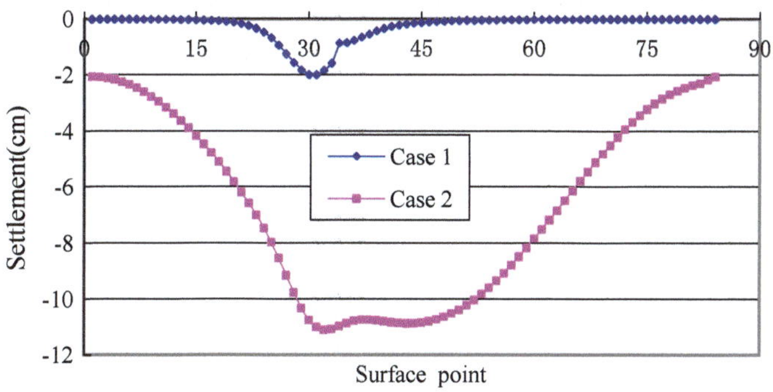

Fig. 4. The surface settlement change curves of two work conditions

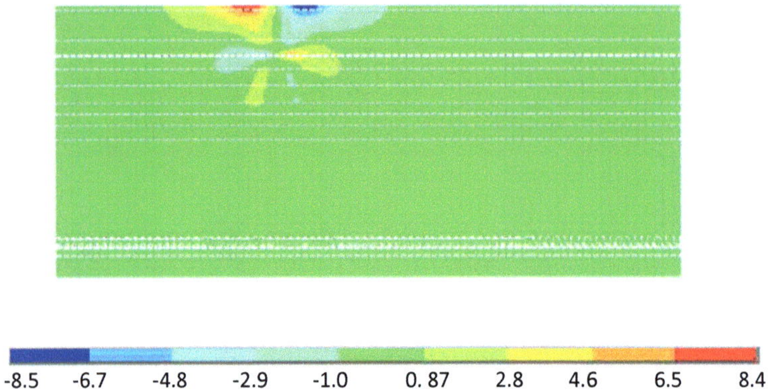

Fig. 5. Horizontal displacement nephogram of model 1 in case 1 (unit: mm)

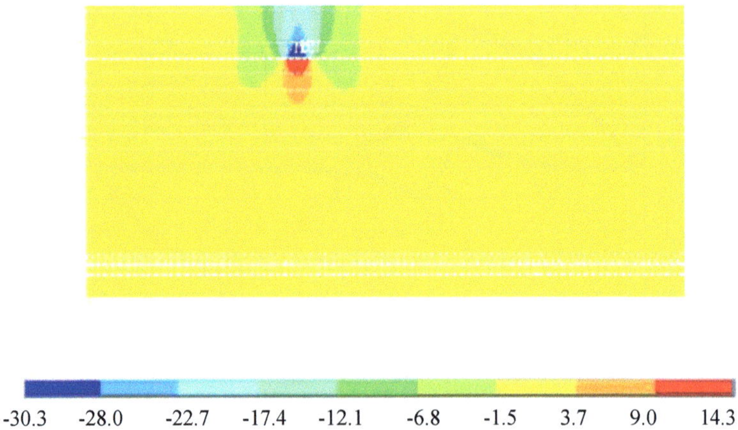

Fig. 6. Settlement nephogram of model 1 in case 1 (unit: mm)

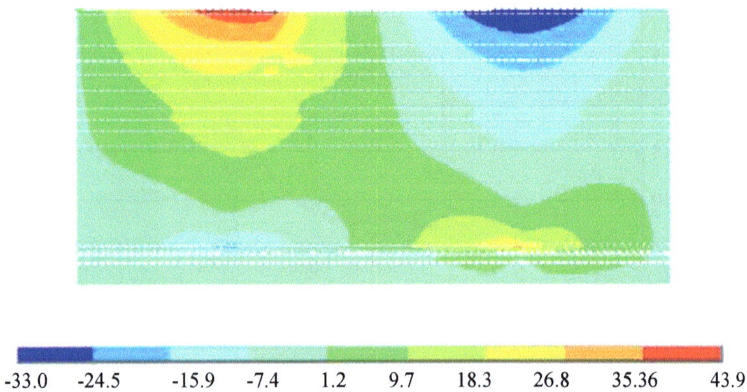

Fig. 7. Horizontal displacement nephogram of model 2 (unit: mm)

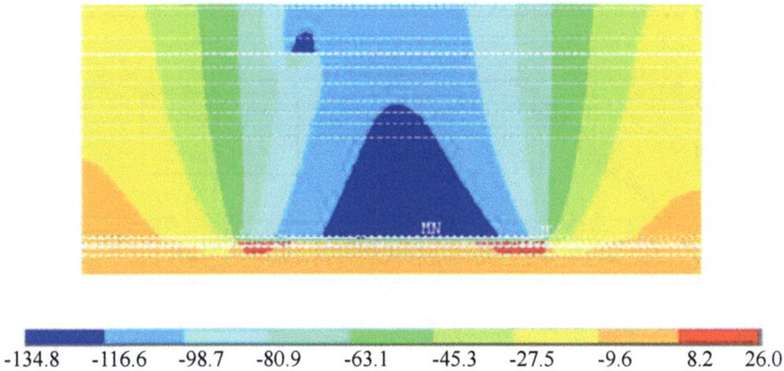

Fig. 8. Settlement nephogram of model 2 (unit: mm)

2.4 Model 2

2.4.1 Calculation and Boundary Conditions

The finite element mesh of model 2 is shown in Fig. 9. The length of the model along the vertical direction of the coal seam is 950 m. The model extends downward from the coal seam is 50 m. The model is divided into 2812 units, the nodes numbers are 2793.

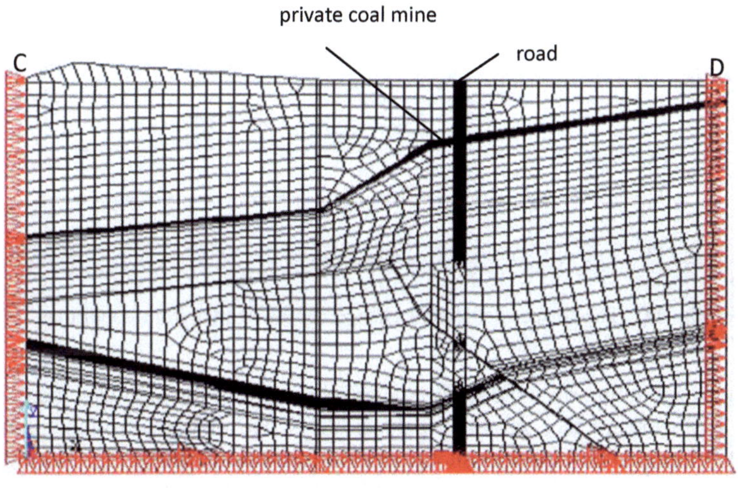

Fig. 9. The boundary condition and element partition of model 2

2.4.2 Calculation Condition

Case 1: The small coal mines has been finished.

Case 2: Large coal mining has been exploited completely.

2.4.3 Calculation Analysis

In accordance to the calculation process, there are 59 nodes on the ground surface from C to D, and ground surface horizontal displacement and vertical settlement situation are shown in Figs. 10 and 11.

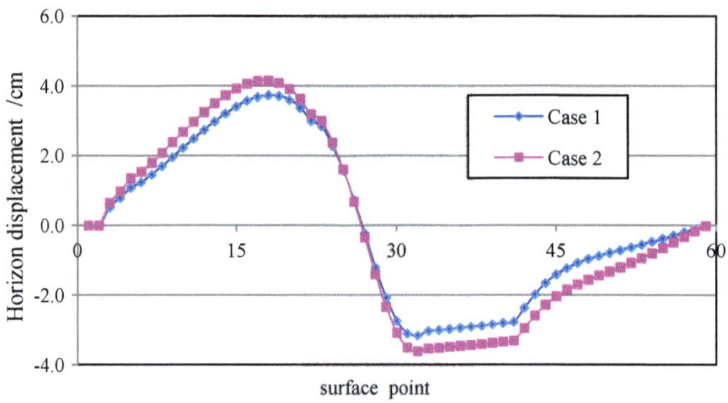

Fig. 10. The surface horizon displacement change curves of two work conditions

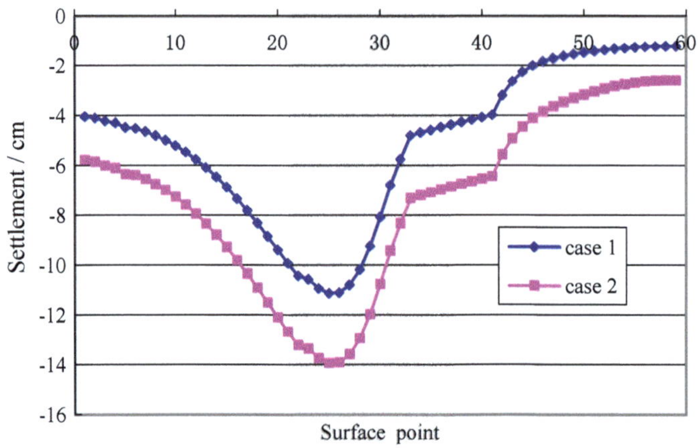

Fig. 11. The surface settlement change curves of two work conditions

This chapter summarizes the surface deformation law of the highway and the village near the 109 working face mining, analyzes the scope of the subsidence basin, and predicts the horizontal displacement and vertical settlement value (Figs. 12 and 13).

The surface horizontal displacement and vertical settlement are shown in Table 2. The calculation results are shown in Figs. 14 and 15.

The maximum vertical subsidence is 13.92 cm after the 109 working face mining completed in model 2, and in the region where we care about (the building roads and village locations), the maximum vertical settlement is expected to be about 7.31 cm, the tilt speed of the sinking region from the southwest to the center of the basin is 0.2 mm/m, the deformation curvature of the surface in the region where we care about is 0.013.

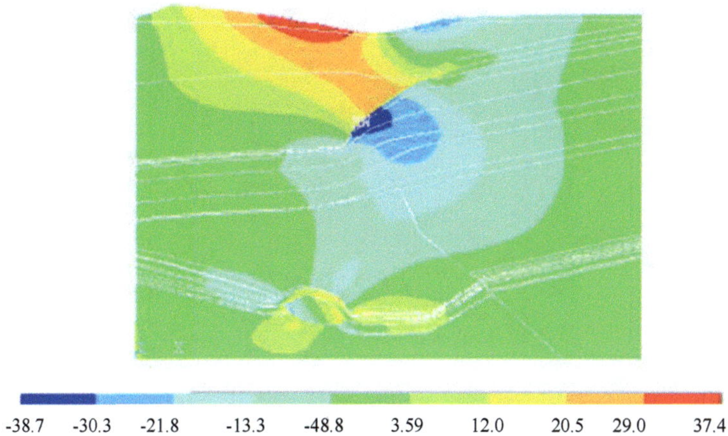

Fig. 12. Horizontal displacement nephogram of model 1 in case 1 (unit: mm)

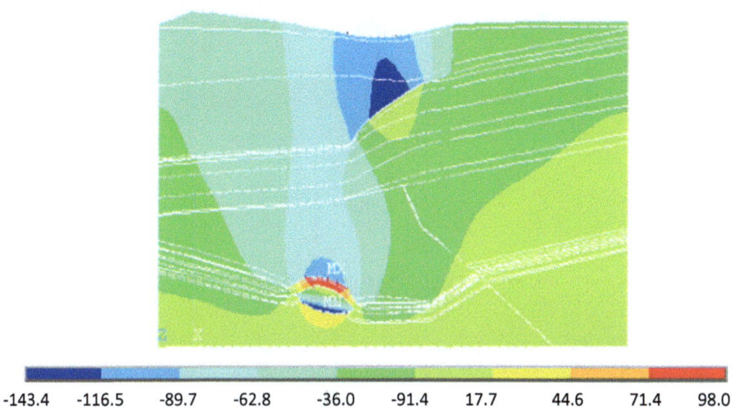

Fig. 13. Settlement nephogram of model 1 in case 1 (unit: mm)

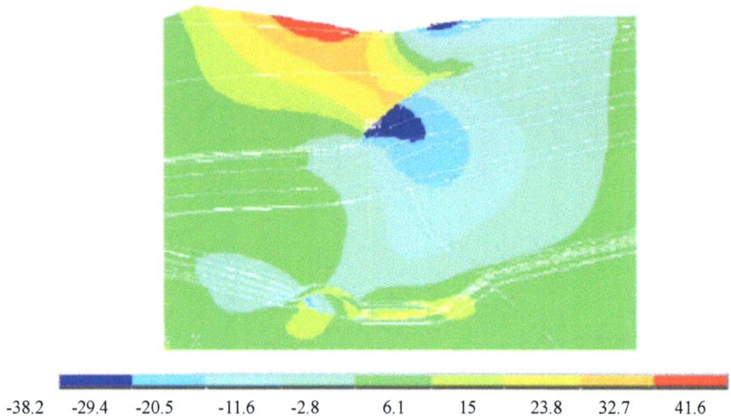

Fig. 14. Horizontal displacement nephogram of model 1 in case 2 (unit: mm)

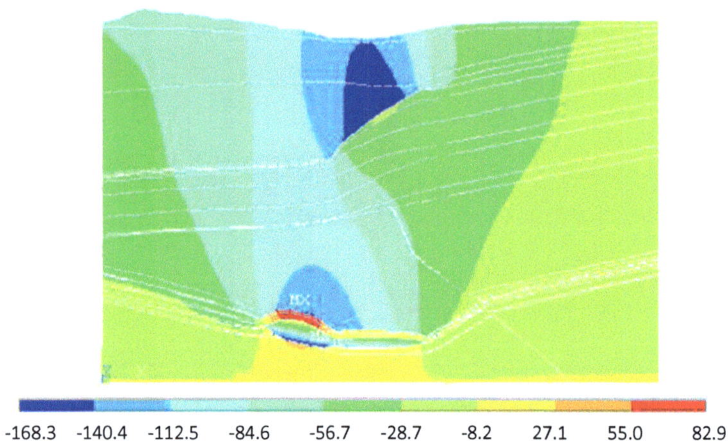

Fig. 15. Settlement nephogram of model 1 in case 2 (unit: mm)

3 Conclusion

In this paper, the numerical simulation is carried out to study the variation laws of the ground surface displacement under different work conditions:

(1) Surface deformation is stacked by the exploitation of two coal mining, the surface damage of the road is mainly caused by the private coal mine exploitation, the long term deformation caused by the large coal mine is not the main influence of the damage of the road and village.
(2) Mining subsidence caused by large coal mining has certain influence on road operation. The additional vertical displacement and horizontal deformation caused by mining are smaller, and the surface deformation can be ignored after the

completion of all the mining faces in the coal mine, and the surface building needn't to be maintenance.

(3) It is suggested to set up the surface deformation monitoring system, strengthen the monitoring, analysis the surface deformation law, to deal with the emergencies.

Acknowledgements. This work was supported by the Youth Innovation Promotion Association CAS, Science and Technology Service Network Initiative (No. KFJ-STS-ZDTP-015) and the Natural Science Foundation of China (No. 41472286, 41472290, 41402276 and 41672312).

References

Yilmaz, E.: Effect of curing under pressure on compressive strength development of cemented paste backfill. Miner. Eng. **22**, 772–785 (2009)

Cooke, R.: Hydraulic backfill distribution systems for deep mines. J. Mines Met. Fuels (9–10): 363–370 (1997)

Wang, Y., Wang C., Ma, W et al.: Feasibility study on cemented backfill in longwall coal mine for surface subsidence control. In: Proceedings of the 6th International Symposium on Mining with Backfill (1998)

Badr, S.A.: Numerical Analysis of coal yield Pillars at deep Longwall Mines. Ph.D.Thesis in preparation. Department of Mining Engineering, Colorado School of Mines, Golden, Colorado (2003)

Sirivardane, H., Amanda J.: Displacement based approach for prediction of subsidence cause by long wall mining using numerical method. In: Proceeding of the 7th International Conference on Computer. Method (1991)

Yao X.L., Reddish, D.J., Whittaker, B.N.: Nonlinear finite element analysis of surface subsidence using from inclined seam extraction. Int. J. RODK Mech. Mining SCL (1993)

Brady B.H.G., Brown E.T.: Mining-Induced Surface Subsidence. Rock Mechanics for Underground Mining, third edition, pp. 484–517 (2004)

Cui, X.M., Miao, X.X., Wang, J.A.: Improved prediction of differential subsidence caused by underground mining. Int. J. Rock Mech. Min. Sci. **37**(4), 615–627 (2000)

Yuan, L.: Monitoring of Huaihe Dike deformation caused by mining. J. China Univ. Min. Technol. **11**(1) (2001)

Cui, X.M., Wang, J., Liu, Y.: Prediction of progressive surface subsidence above long-wall coal mining using a time function. Int. J. Rock Mech. Min. Sci. **38**(7), 1057–1063 (2001)

Moes, N., Belytschko, T.: Extend finite element method for cohesive crack growth. Eng. Fract. Mech. **69**(7):813–833 (2002)

Ferrari, C.R.: Residual coal mining subsidence-somefacts. Min. Technol. **7**(9), 177–183 (1997)

Prus, J.: Influence of mining operation upon the condition of environment in the industrial area of upper. In: Proceeding of Second World Mining Environment Congress, pp. 9–52 (1997)

Yu, G., Xie, H., Zhan, J. et al.: Fractal evolution crack network in overburden rock. Discrete Dyn. Nat. Soc. **35**(8):1107–1111 (1998)

Settlement of Composite Foundation with Sparse PTC (Pre-stressed Tubular Concrete) Capped-Piles Under Embankment

Bruce Zhi-Feng Wang[1(✉)], Ya-Qiong Wang[1], and Jason Wen-Chieh Cheng[2]

[1] Department of Geotechnical and Tunnelling Engineering, School of Highway, Chang'an University, Xi'an, Shaanxi 710064, China
zhifeng.wang@chd.edu.cn, wangyq@126.com
[2] Institute of Tunnel and Underground Structure Engineering, School of Civil Engineering, Xi'an University of Architecture and Technology, Xi'an, 710055, China
w-c.cheng@xauat.edu.cn

Abstract. This paper presents an investigation into the settlement of the composite foundation with sparse PTC (pre-stressed tubular concrete) capped-piles under embankment by theory analysis. Based on the assumption that the distribution of skin frictions on PTC capped-piles is simplified as two triangles, a method for calculating the additional stresses of the composite foundation with sparse PTC capped-piles is proposed. By the combination of Mindlin-Geddes' solution and Boussinesq's solution, the formula of additional stresses of single PTC capped-pile foundation is derived by considering the radius of influence range. A case history of highway embankment construction conducted in Jiangsu province, which involved installation of sparse PTC capped-piles for soil reinforcement, was briefly introduced and analyzed, and the observed settlements after completion of the embankment construction reached to 428 mm. The proposed method and the layer-wise summation method were adopted to calculate the settlement of the composite foundation with sparse PTC capped-piles. The comparison results indicate that the calculated settlement of the composite foundation have a good agreement with the field data. The analysis results also suggest that, when the diameter of influence range is eight times larger than the diameter of the pile, the additional stress field shows no significant variation with increasing the influence range.

1 Introduction

With the development of construction of highway infrastructure in China, the composite foundation with sparse rigid-piles has been widely used for controlling the ground settlement (Poulos 2001; Liu et al. 2007). Many researchers studied the behavior of the pile-supported embankments by conducting laboratory tests and field tests (Poulos 2007; Manna and Baidya 2009; Hong et al. 2014). Okyay et al. (2014)

S. Wang et al. (eds.), *New Solutions for Challenges in Applications of New Materials and Geotechnical Issues*, Sustainable Civil Infrastructures,
https://doi.org/10.1007/978-3-319-95744-9_15

investigated the behavior of a granular earth-platform with rigid pile reinforcement by the centrifuge model tests. Chen et al. (2010) analyzed the settlements, earth pressures, and pore-water pressures during the construction pile-supported embankments by the field tests. Generally, the calculation of additional stress and settlement of composite foundation with sparse PTC (pre-stressed tubular concrete) capped-piles under embankment is a key point in the design stage for the geotechnical engineers (Chen et al. 2008; Sloan et al. 2013; Al-Ani et al. 2014). In order to obtain a reasonable calculation result of the additional stress of composite foundation with sparse PTC capped-piles, the distribution of skin frictions need to be considered in a reasonable way (Liu et al. 2013; Yu et al. 2014). Presently, the finite element method is usually adopted to investigate the distribution of skin frictions (Pham et al. 2004; Liu et al. 2010; Girout et al. 2014), indicating that the skin frictions increase to the maximum value with the increase of the depth from the top of the pile to a particular location of the pile, and then decrease with the increase of the depth from the particular location of the pile to the bottom of the pile. Published literatures (Jardine et al. 1995; Zhang 1999; Wang et al. 2006; Anyaegbunam 2014) suggest that a reasonable result for the analysis of rigid piles in homogeneous soils may be acquired by using the Geddes's solution (Geddes 1966). Using the Geddes's solution and the displacement coordination theory, Huang and Jia (1983) proposed a method for calculating the settlement of pile foundation, but there is some discrepancy between the calculated data and measured data for the case of composite foundation with sparse PTC capped-piles. Therefore, there need to develop a simple method for calculating the additional stresses and settlement of the composite foundation with sparse PTC capped-piles under embankment.

In the present study, combining the Mindlin-Geddes' solution and Boussinesq's solution (Boussinesq 1885; Mindlin 1936; Geddes 1966), the formula of additional stresses of single PTC capped-pile foundation is derived by considering the radius of influence range. A method for calculating the additional stresses of the composite foundation with sparse PTC capped-piles is proposed on the basis of the assumption that the distribution of skin frictions on PTC capped-piles can be simplified as two triangles, and a case history is introduced to demonstrate its applicability.

2 Assumption for Distribution of Skin Frictions

A number of laboratory and numerical investigations on the distribution of skin frictions of single piles indicate that the skin frictions increase to the maximum value with the increase of the depth from the top of the pile to a particular location of the pile, and then decrease with the increase of the depth from the particular location of the pile to the bottom of the pile (Zhang et al. 2011). In this study, as shown in Fig. 1, the distribution of skin frictions of single PTC pile can be assumed as: (*i*) the value of the skin frictions in the depth of H_1 is the maximum value ($q_{\max}$); (*ii*) the values of the skin frictions in the top and bottom of the pile are the minimum values.

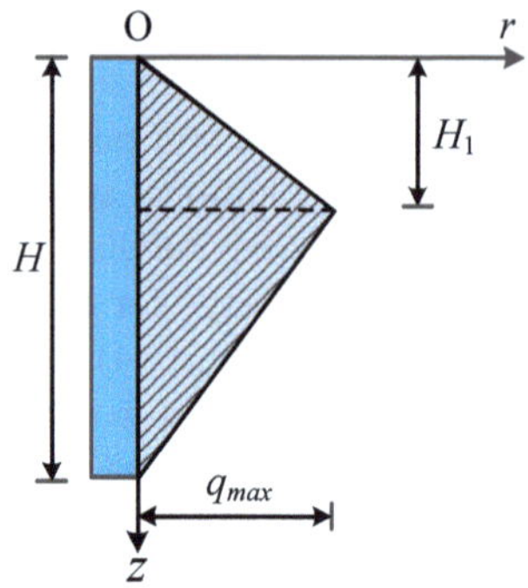

Fig. 1. Illustration of skin frictions distribution of piles

3 Calculation Model for Single PTC Capped-Pile Under Embankment

Engineering practice of the composite foundation with sparse rigid piles shows that the soil with a distance of about $6D$ (D is the diameter of the pile) around a single rigid pile can be influenced (Motta 2013; Faro et al. 2015). In general, the pile spacing is within the range of $7.5D$–$9D$ for the composite foundation with sparse rigid piles, and many numerical analyses investigating the behavior of the composite foundation with sparse rigid piles suggest that the pile group effect is not obvious in this case (Higgins et al. 2013; Song and Xu 2014).

Figure 2 depicts the calculation model for single PTC capped-pile under embankment. As can be seen, for simplification, a single PTC capped-pile foundation under the cushion with a uniformly distributed load is established to study its behavior. The thicknesses of the cap and the substratum are H_c and H_s respectively, and the length of the PTC pile is H. Commonly, Geddes's solution is adopted to obtain the additional stress due to the variation of stress state of the soil. In this study, the calculation process can be divided into two issues: (*i*) for the upper part of the pile ($H/3$), it can be directly calculated by Geddes's solution; (*ii*) for the lower part of the pile ($2H/3$), it can be calculated by subtraction of rectangle and triangle, as indicated in Fig. 3.

As shown in Fig. 4, an influence range is assumed to calculate the additional stress caused by the variation of stress state of the pile. In order to obtain the additional Boussinesq's stress due to the variation of stress state of the soil between piles, the calculation area can be divided into two aspects: (*i*) for the part outside the influence range, it can be directly calculated by the strip load (P); (*ii*) for the part inside the influence range, it can be calculated by subtraction of the circular area and the cap areas, as indicated in Fig. 5.

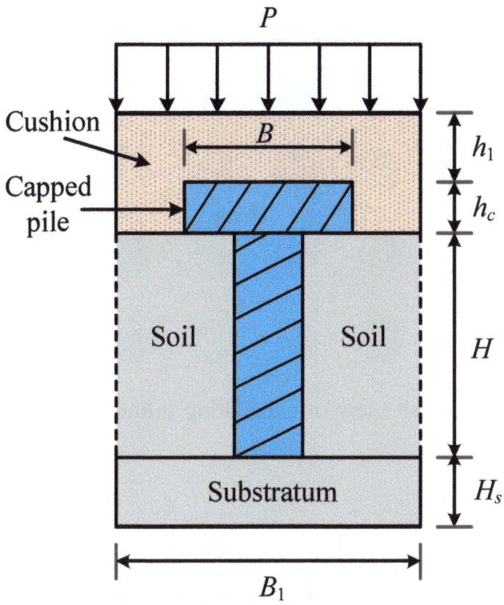

Fig. 2. Calculation model of PTC capped-pile foundation

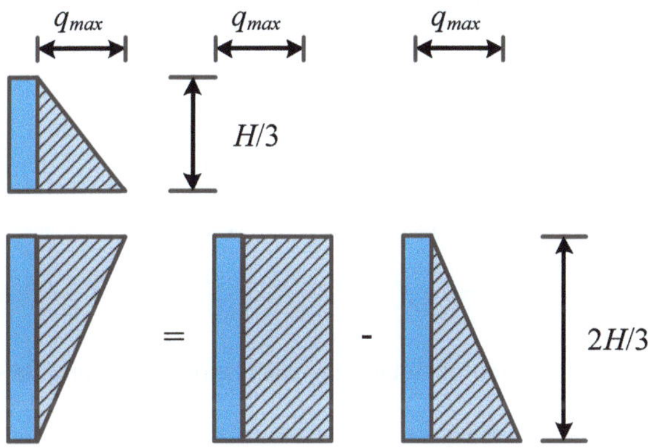

Fig. 3. Illustration of simplifying calculation

4 Additional Stress for Composite Foundation with PTC Capped-Piles

It is assumed that the skin frictions at the depth of $H/3$ reach the maximum value ($q_{\max}$), and the following equation can be obtained:

$$2\pi r_0 \frac{Hq_{\max}}{2} = Q(1 - \alpha) \tag{1}$$

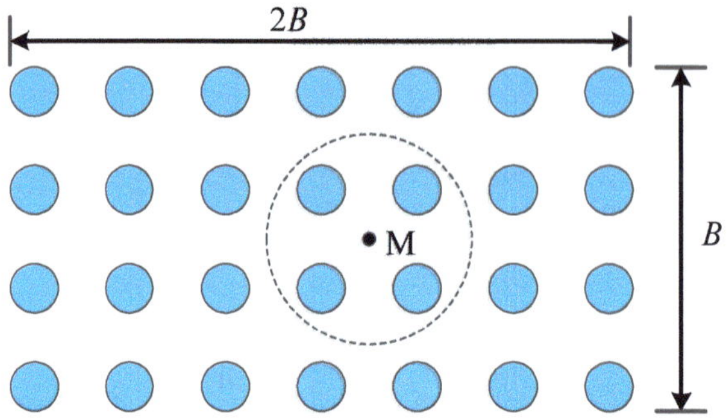

Fig. 4. Plan view of calculating influencing radius

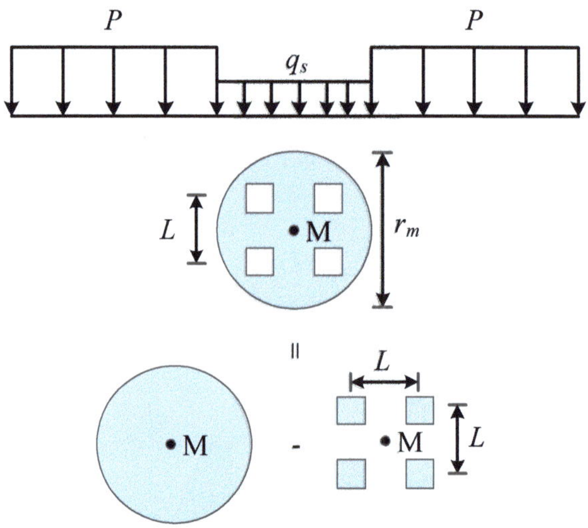

Fig. 5. Illustration of simplifying calculation of soil-pile stress

for the upper part of the pile:

$$\begin{cases} \alpha_1 = 0 \\ \beta_1 = 0 \\ (1 - \alpha - \beta)_1 = (1 - \alpha)/3 \end{cases} \tag{2}$$

for the lower part of the pile:

$$\begin{cases} \alpha_2 = \alpha \\ \beta_2 = 4(1 - \alpha)/3 \\ (1 - \alpha - \beta)_2 = 2(1 - \alpha)/3 \end{cases} \tag{3}$$

where α is the load percentage at the pile end; Q is the load at the top of pile; r_0 is the radius of the pile; H is the length of the pile. It is assumed that the value of strip load is P, and the load percentage at the pile end is N, then:

$$P_p = QL^2N \tag{4}$$

$$q_s = \frac{Q - P_{\mathrm{p}}}{L^2 - l^2} \tag{5}$$

where P_{p} is the load for piles; q_{s} is the load for soils; L is the pile spacing; l is the length of the cap.

The additional stress caused by the load for piles can be calculated as follows:

$$\sigma_{Pi} = \frac{Q}{\left(H/3\right)^2}(1-\alpha-\beta)_{i1}I_{it1} + \frac{Q}{\left(2H/3\right)^2}\left[\alpha_{i2}I_{ib2} + \beta_{i2}I_{ir2} + (1-\alpha-\beta)_{i2}I_{it2}\right] \tag{6}$$

$$\sigma_P = \sum_{i=1}^{n} \sigma_{Pi} \tag{7}$$

where n is the number of the piles in the influence range.

The additional stress caused by the load for soils can be calculated as follows:

$$\sigma_S = \sigma_{S1} + \sigma_{S2} \tag{8}$$

$$\sigma_{S2} = \sigma'_S - n\sigma'_p \tag{9}$$

$$\sigma_{S1} = p(\alpha_c I_s + \alpha_c II_s + \alpha_c III_s + \alpha_c IV_s) \tag{10}$$

$$\sigma'_S = q_s\left[1 - \frac{z^3}{\left(r_0^2 + z^2\right)^{3/2}}\right] \tag{11}$$

$$\sigma'_P = p(\alpha_c I_p - \alpha_c II_p - \alpha_c III_p + \alpha_c IV_p) \tag{12}$$

where σ_{S1} is the additional stress induced by the load for the soil outside the influence range; σ_{S2} is the additional stress induced by the load for the soil inside the influence range; σ'_S is the additional stress induced by the load for the soil which has the same area with the influence range; σ'_p is the additional stress induced by the load for the soil which has the same area with the caps; $\alpha_c I_s \sim \alpha_c IV_s$ and $\alpha_c I_p \sim \alpha_c IV_p$ are the coefficients for calculating the values of σ'_S and σ'_p using corner point method, as indicated in Fig. 6.

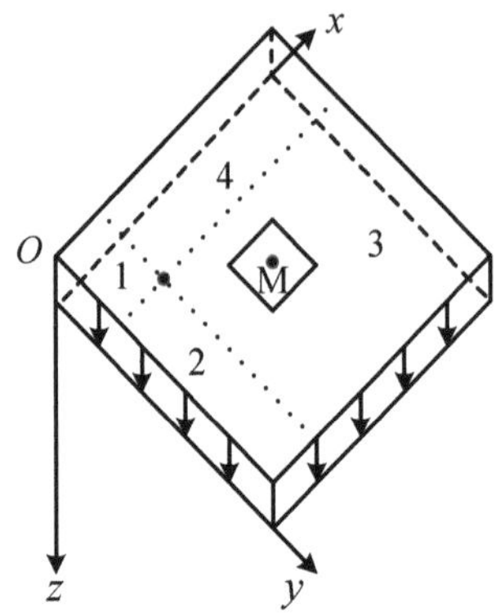

Fig. 6. Illustration of calculating additional stress of point M

5 Analysis of Case History Conducted in Jiangsu Province, China

In Jiangsu province, the PTC piles were commonly adopted for soil reinforcement under the highway embankment. In this section, a case history conducted in Lianyungang city of Jiangsu province was analyzed. Figure 7 shows the physical and mechanical properties of subsoils, and the construction site mainly contains the clay, soft silty clay, and the sand. The thickness of the clay and soft silty clay is about 4 m and 21 m respectively. The soft silty clay has a high water content of 45% and a low constrained modulus of 3.2 MPa. Figure 8 presents the sectional and plan view of the layout of PTC piles and settlement observations. As can be seen, the height of the embankment is about 5.7 m, and the width of the embankment is about 46 m. The length of the pile is about 25 m, and the diameter of the pile is about 0.4 m. The PTC piles are arranged in the form of square, and the pile spacing is about 3.5 m. The embankment was filled to the designed height of 5.7 m in a period about 50 days. Two settlement observations (O1 and O2) were installed to monitor the settlement after the beginning of the embankment construction. The observation lasted ten months after the construction.

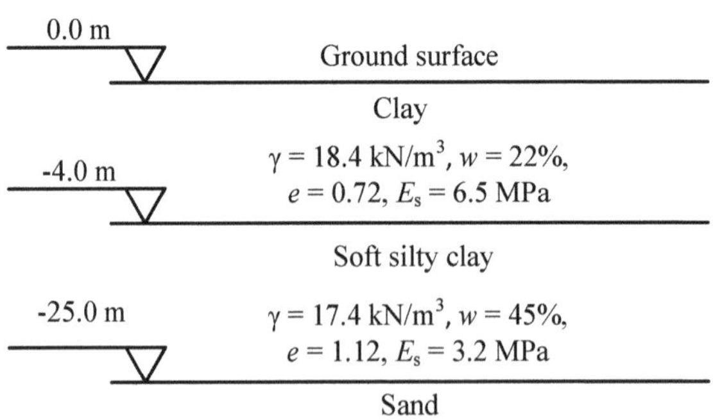

Fig. 7. Physical and mechanical properties of subsoils: w, natural water content of soil; γ, unit weight of soil; E_s, constrained modulus of soil; e, void ratio of soil

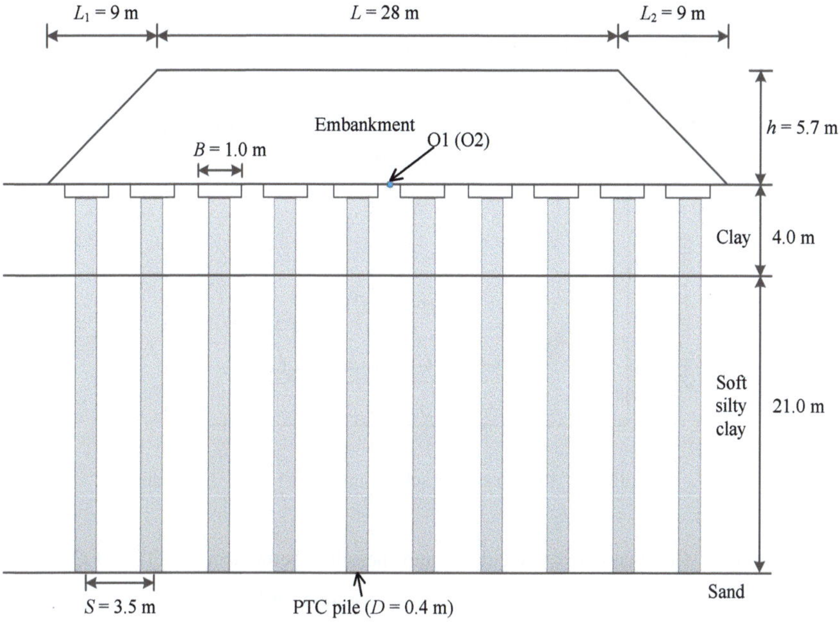

(a) Sectional view

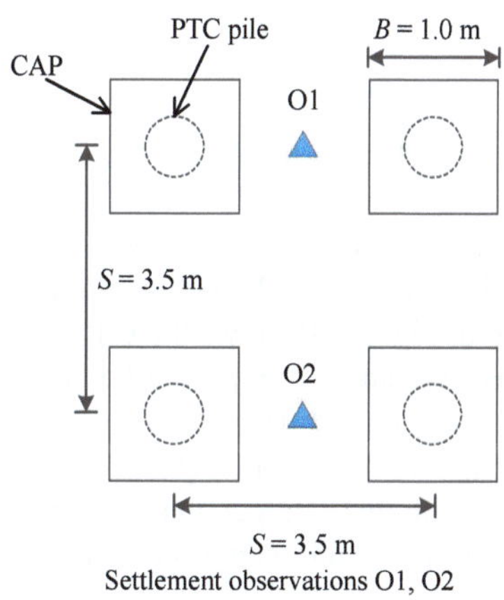

(b) Plan view

Fig. 8. Sectional and plan view of the layout of PTC piles and settlement observations

Figure 9 shows the observed curve of the settlement with the construction time. As can be seen, the settlement increased with the increase of the fill height of embankment. Figure 9 also indicates that the maximum settlement of O1 and O2 are 405 mm and 428 mm, respectively. By the combination of the proposed method and the layer-wise summation method, the calculated settlement of the composite foundation with sparse PTC capped-piles is about 420 mm, as indicated in Table 1. It can be seen that the calculation error between the predicted values and the field data is

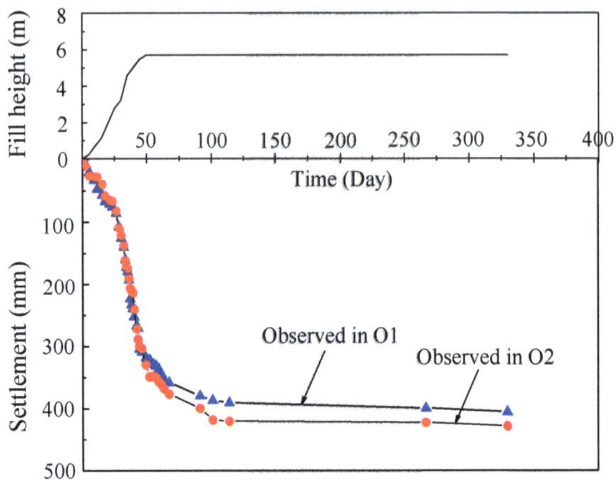

Fig. 9. Variation of the observed settlements with time in O1 and O2

Table 1. Comparisons of calculated and measured settlements in O1 and O2

Settlement observations	Measured settlement (mm)	Calculated settlement (mm)	Calculation error (%)
O1	405	420	3.6
O2	428	420	1.9

about 3.6% and 1.9% respectively, which demonstrate the applicability of the method proposed in this study.

Using the proposed method in this study, the additional stress for composite foundation with the different influence ranges (the radius of influence range is assumed as: 8D, 14D and 19D) is analyzed. Figure 10 shows the calculated results for the cases of the different influence ranges. As can be seen, the difference among the calculated additional stress for these three influence ranges is not obvious, and the maximum values are 114.6 kPa, 112.5 kPa, and 107.5 kPa, respectively. Figure 10 also indicates that, when the pile spacing is eight times larger than the diameter of the pile, the additional stress field shows no significant variation with increasing the influence range.

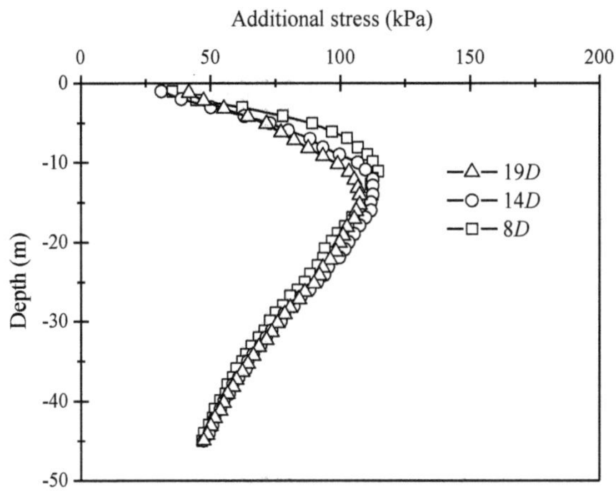

Fig. 10. Variation of additional stress with depth for different influence ranges

6 Conclusions

The additional stresses of the composite foundation with sparse PTC (pre-stressed tubular concrete) capped-piles under embankment are investigated in this study. Based on the assumption that the distribution of skin frictions on PTC capped-piles is simplified as two triangles, a method for calculating the additional stresses of the composite foundation with sparse PTC capped-piles is proposed. By the combination of Mindlin-Geddes' solution and Boussinesq's solution, the formula of additional stresses of single PTC capped-pile foundation is derived by considering the radius of influence range. A case history involving installation of sparse PTC capped-piles for soil reinforcement is introduced and analyzed, and the observed settlements after completion of the embankment construction reached to 428 mm. The proposed method and the layer-wise summation method were used to calculate the settlement of this case, and the comparison results indicate that the calculated data by the proposed method have a good agreement with the field data. Analysis results suggest the difference among the calculated additional stress for these three influence ranges is not obvious, and the maximum values are 114.6 kPa, 112.5 kPa, and 107.5 kPa respectively. The comparison result also indicates that, when the pile spacing is eight times larger than the diameter of the pile, the additional stress field shows no significant variation with increasing the influence range. It is suggested that the method may be a useful tool for the design of soft subsoil improvement resulting from PTC piles.

Acknowledgments. The research described in this study was funded by the National Nature Science Foundation of China (NSFC) (Grant No. 41702287), the Fundamental Research Funds for the Central Universities (Grant No. 310821161022 and 310821153312), and China Postdoctoral Science Foundation (2015M570803 and 2016T90877). These financial supports are gratefully acknowledged.

References

Al-Ani, W., Wanatowski, D., Chan, S.: Numerical analysis of piled embankments on soft soils. In: Proceedings of Geo-Shanghai 2014: Ground Improvement and Geosynthetics, Shanghai (2014)

Anyaegbunam, A.: Complete stresses and displacements in a cross-anisotropic half-space caused by a surface vertical point load. Int. J. Geomech. (2014). https://doi.org/10.1061/(ASCE)GM.1943-5622.0000260

Boussinesq, J.: Application des potentiels a l'etude de l'equilibre et du mouvement des solides elastiques. Gauthier-Villars, Paris, France (1885)

Chen, R.P., Chen, Y.M., Han, J., Xu, Z.Z.: A theoretical solution for pile-supported embankments on soft soil under one dimensional compression. Can. Geotech. J. (2008). https://doi.org/10.1139/T08-003

Chen, R.P., Xu, Z.Z., Chen, Y.M., Ling, D.S., Zhu, B.: Field tests on pile-supported embankments over soft ground. J. Geotech. Geoenviron. Eng. (2010). https://doi.org/10.1061/(ASCE)GT.1943-5606.0000295

Faro, V., Consoli, N., Schnaid, F., Thomé, A., da Silva Lopes, L.: Field tests on laterally loaded rigid piles in cement treated soils. J. Geotech. Geoenviron. Eng. (2015). https://doi.org/10.1061/(ASCE)GT.1943-5606.0001296

Geddes, J.D.: Stress in foundation soils due to vertical subsurface loading. Géotechnique (1966). https://doi.org/10.1680/geot.1966.16.3.231

Girout, R., Blanc, M., Dias, D., Thorel, L.: Numerical analysis of a geosynthetic-reinforced piled load transfer platform-validation on centrifuge test. Geotext. Geomembr. (2014). https://doi.org/10.1016/j.geotexmem.2014.07.012

Higgins, W., Vasquez, C., Basu, D., Griffiths, D.: Elastic solutions for laterally loaded piles. J. Geotech. Geoenviron. Eng. (2013). https://doi.org/10.1061/(ASCE)GT.1943-5606.0000828

Hong, W., Lee, J., Hong, S.: Full-scale tests on embankments founded on piled beams. J. Geotech. Geoenviron. Eng. (2014). https://doi.org/10.1061/(ASCE)GT.1943-5606.0001145

Huang, S.M., Jia, Z.Y.: Calculation on settlement of pile foundation in soft soil. In: Proceedings of 4th National Symposium on Soil Mechanics and Foundation Engineering. Wuhan, July (1983) (in Chinese)

Jardine, R.J., Lehane, B.M., Smith, P.R., Gildea, P.A.: Vertical loading experiments on rigid pad foundations at Bothkennar. Géotechnique (1995). https://doi.org/10.1680/geot.1995.45.4.573

Liu, H.L., Ng, C.W.W., Fei, K.: Performance of a geogrid—reinforced and pile-supported highway embankment over soft clay: case study. J. Geotech. Geoenviron. Eng. https://doi.org/10.1061/(asce)1090-0241(2007)133:12(1483) (2007)

Liu, H.L., Kong, G.Q., Ding, X.M., Chen, Y.M.: Performances of large-diameter cast-in-place concrete pipe piles and pile groups under lateral loads. J. Perform. Constructed Facil. (2013). https://doi.org/10.1061/(ASCE)CF.1943-5509.0000304

Liu, L., Zheng, G., Han, J.: Numerical analysis of lateral behavior of rigid piles to support embankments. In: Proceedings of Geo-Shanghai 2010: Ground Improvement and Geosynthetics, Shanghai, June (2010)

Manna, B., Baidya, D.K.: Vertical vibration of full-scale pile-analytical and experimental study. J. Geotech. Geoenviron. Eng. (2009). https://doi.org/10.1061/(ASCE)GT.1943-5606.0000110

Mindlin, R.D.: Forces at a point in the interior of a semi-infinite solid. J. Appl. Phys. **9**(5), 195–202 (1936)

Motta, E.: Lateral deflection of horizontally loaded rigid piles in elastoplastic medium. J. Geotech. Geoenviron. Eng. (2013). https://doi.org/10.1061/(ASCE)GT.1943-5606.0000771

Okyay, U.S., Dias, D., Thorel, L., Rault, G.: Centrifuge modeling of a pile-supported granular earth-platform. J. Geotech. Geoenviron. Eng. (2014). https://doi.org/10.1061/(ASCE)GT.1943-5606.0001004

Pham, H., Suleiman, M.T., White, D.J.: Numerical analysis of geosynthetic-rammed aggregate pier supported embankment. In: Proceedings of Geo-Trans 2004 Conference: Geotechnical Engineering for Transportation Projects, Los Angeles, July (2004)

Poulos, H.G.: Piled raft foundations: design and applications. Géotechnique (2001). https://doi.org/10.1680/geot.2001.51.2.95

Poulos, H.G.: Design charts for piles supporting embankments on soft clay. J. Geotech. Geoenviron. Eng. (2007). https://doi.org/10.1061/(ASCE)1090-0241(2007)133:5(493)

Sloan, J.A., Filz, G.M., Collin, J.G.: Field-scale column-supported embankment test facility. Geotech. Test. J. (2013). https://doi.org/10.1520/GTJ20120229

Song, S., Xu, C.: Numerical analyses of soil arching in rigid pile supported embankments. In: Proceedings of Geo-Shanghai 2014: Ground Improvement and Geosynthetics, Shanghai, May (2014)

Wang, C., Pan, E., Tzeng, C., Han, F., Liao, J.: Displacements and stresses due to a uniform vertical circular load in an inhomogeneous cross-anisotropic half-space. Int. J. Geomech. (2006). https://doi.org/10.1061/(ASCE)1532-3641(2006)6:1(1)

Yu, F., Wu, C., Liu, H., Xia, T.: Evaluation of the settlement of embankment supported by cast-in-place concrete tubular piles. In: Proceedings of Geo-Hubei 2014: Characterization, Modeling, and Evaluation of Geotechnical Engineering Systems, Yichang, July (2014)

Zhang, C., White, D., Randolph, M.: Centrifuge modeling of the cyclic lateral response of a rigid pile in soft clay. J. Geotech. and Geoenviron. Eng. (2011). https://doi.org/10.1061/(ASCE)GT.1943-5606.0000482

Zhang, L.M.: Settlement patterns of soft soil foundations under embankments. Can. Geotech. J. (1999). https://doi.org/10.1139/t99-031

Ground Motion Amplification Induced by Shallow Circular Tunnel in Soft Soil

Mohsen Mousivand[1(✉)], Hesam Aminpour[2], and Nura Ebrahimi[3]

[1] Islamic Azad University, Gonbad kavoos branch, Tehran, Iran
moosivand63@gmail.com
[2] Department of Civil Engineering, Bu-Ali Sina University, Hamedan, Iran
hesam.aminpour1364@yahoo.com
[3] Islamic Azad University, Gonbad Kavoos Branch, Gonbad Kavoos, Golestan, Iran
nura.ebrahimi@gmail.com

Abstract. In recent years, there has been an increasing interest in tunnel excavation to resolve traffic problems especially in populated urban areas. However, tunnel excavations could seriously affect ground surface behavior static and dynamic load conditions. Several studies have estimated the ground motion induced by tunnels in homogeneous, isotropic and linear elastic soils, however there is still insufficient data on behavior in real conditions such as non-linear and anisotropic behavior of soil in response to tunnel excavation. The aim of this paper is to examine the ground motion amplification due to excavation of a circular tunnel in shallow depth in soft soil. To this aim, at first, using a verified finite element code, effective parameters of problem including depth and ductility of tunnel, frequency content and soil material set are investigated. Then, the effects of underground circular structure on dynamic response of ground are assessed according to real time history earthquakes. Finally the results are compared with green field condition. The results of this study indicate that underground circular structures have significant effects on ground motion amplification. These effects can be twice as large as green field conditions.

Keywords: Tunnel · Finite element code · Soft soil · Shallow depth
Dynamic analysis

1 Introduction

For several decades, due to increasing growth of population of the cities and urgent need for better transportation services, tunnel construction has been recognized as a key strategy for reducing traffic jams and facilitating transportation. In this regard, extensive studies have been conducted by researchers on the effects of tunnel excavation, tunnel stability, and interaction of tunnel structures. In the majority of studies on predicting the ground

---

The original version of this chapter was revised: The missing authors' names were updated.
The correction to this chapter is available at https://doi.org/10.1007/978-3-319-95744-9_18

S. Wang et al. (eds.), *New Solutions for Challenges in Applications of New Materials and Geotechnical Issues*, Sustainable Civil Infrastructures,
https://doi.org/10.1007/978-3-319-95744-9_16

displacement around the tunnels, experimental, numerical, and analytical methods have been mostly utilized. Peck (1969) conducted a field study on a number of circular tunnels excavated in different soils and showed that the subsidence profile at the ground surface was a Gaussian curve. Other studies in this regard have been conducted based on empirical relationships and analytical methods. In analytical methods based on elasticity theories, the surface subsidence profile of the ground could be predicted by equilibrium equations. Experimental studies are based on Peck's equation with only some minor changes in the equation parameters (Attewell and Farmer 1974; Atkinson and Potts 1977; Clough and Schmidt 1981; Mail et al. 1983; Leach 1985; Oteo and Sagaseta 1996). Ground acceleration is one of the important factors for the seismic classification of areas; thus, the effect of underground structures on ground acceleration is very important. Recently, the effect of underground structures on the seismic response of the above-ground structures has been investigated which have shown that the underground structures have a direct impact on the seismic response of superstructures (Yiouta-Mitra et al. 2007). Many strategies have been used to describe the effects of underground cavities on superstructures under the surface and volume waves; however, they are based on simple assumptions (Pao et al. 1973; Dravinski 1983; Wong et al. 1985; Lee 1988; Smerzini et al. 2009). Sun and Wang (2012) studied the ground acceleration for cases with and without tunnel and reported that tunnel would cause a change in the ground acceleration. In another study, Baziar et al. (2014) used laboratory and software models to investigate the effects of rectangular tunnel on the ground acceleration; they applied the records of sinusoid acceleration and real earthquake to the soil floor and stated that the rectangular tunnel reduced and increased the acceleration spectrum in short and long periods, respectively. Also, the construction of rectangular tunnel increased maximum earthquake acceleration compared to the tunnel-less case; since records of real earthquakes are different from each other, effects of rectangular tunnel on the earthquake acceleration record would be also different (Baziar et al. 2014). Cilingir and Madabhushi (2011) investigated the effects of earthquake forces with different frequencies and amplitudes on the behavior of circular and square tunnels using ABAQUS software and reported that, with the increase in frequency, Fourier spectrum and acceleration were reduced and increased in shorter and longer periods, respectively. Also, increase in the earthquake amplitude would increase the axial forces and bending moment of tunnel wall. Abuhajar et al. (2011) also modeled a square tunnel in sandy soil of Nevada to study the effects of tunnel excavation on the earthquake acceleration by modeling different records and stated that the soil density would influence the acceleration amplitude of earthquakes. They also found that acceleration amplitude is a very important factor in determining reduced earthquake ground acceleration; i.e. increase in the acceleration amplitude would increase the earthquake ground deceleration. Maleki et al. (2011) evaluated the effect of circular tunnels excavation on adjacent structure displacement with 2D Plaxis. They reported that, tunnel excavation-induced ground movement is strongly affected by stiffness of adjacent structure. Besharat et al. (2012) studied the effect of underground structures on the ground surface during the earthquake. Using a case study of Niayesh-Sadr tunnel, they stated that the excavation increases the earthquake ground acceleration above the tunnel. The stresses around the tunnel also increased. Pitilakis et al. (2014) studied the effects of above ground structures on circular tunnels. Their Studies showed that above ground structures may cause an increase in the tunnel dynamic response which is more evident for stiff tunnels in shallow burial depths. The main and primary aim of the present study is to examine the effect of tunnel excavation on dynamic properties of soil and its impact on the response of adjacent structures placed near the tunnel. For this purpose, large numbers of numerical simulations were conducted and in these analyses the recorded acceleration of the ground surface was compared for the cases with and without tunnel. Also, using Seismosignal software, several characteristics of the earthquake record including response spectrum, Fourier spectrum, and effective duration of motion were studied in the cases with and without tunnel.

1.1 Modeling and Parameters

The numerical simulation and analyses in this study are carried out by PLAXIS finite element (FE) software. A plane strain condition is considered for all the analyses and soil was assumed to be only composed of one layer without underground water. The Mohr-Coulomb constitutive soil model, which is an elastic-perfectly plastic model, is selected to simulate the behavior of soil. The shear strength parameters of soil are tabulated in Table 1. In this study, the tunnel was excavated in a circular shape with constant area and depth of ($\frac{d}{h} = 0.64$) in the soil layer, as shown in Fig. 1, and the concrete lining was used for the tunnel wall. The mechanical properties of tunnel lining are listed in Table 2.

Table 1. Soil physical properties (Afifipour et al. 2011).

H	L	γ_{sat}	γ_d	E	ψ	$\emptyset$	C	R_{inter}	ϑ	α	β
m	m	$KN/_{m^2}$	$KN/_{m^2}$	$KN/_{m^2}$	Degree	Degree	$KN/_{m^2}$				
50	200	17	17	5E4	5	29	0.4	0.7	0.3	0.01	0.001

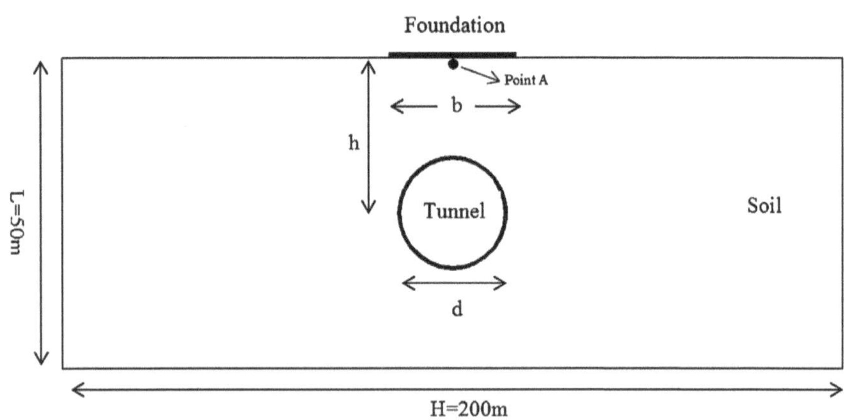

Fig. 1. Soil and tunnel dimensions.

Table 2. Mechanical parameters of tunnel lining.

β	α	ν	EI	EA	d
0.001	0.01	0.25	$KN \cdot m^2/_m$	$KN/_m$	m
			8.218E4	8.05E6	0.35

For modeling, the foundation bending element with an elastic behavior was adopted. The modeled concrete foundation was of apron type with the width of 14 m and thickness of 0.7 m whose characteristics are presented in Table 3. Also, compressive strength of the concrete was considered and the concrete specific weight was assumed to be 24,000.

Table 3. Parameters of structure foundation (Maleki et al. 2011).

EA	EI	ν	α	β
$KN/_m$	$KN \cdot m^2/_m$	0.25	0.01	0.001
1.75E7	7.145E5			

Fixed and energy absorbing boundary conditions were used. In the fixed boundary conditions, a roller support was considered for vertical lines of the soil mass and a hinge was considered for the horizontal line of the soil mass floor; also, no support was assumed for the horizontal line above soil mass; therefore, the soil could only move in a vertical line with no horizontal movements. During the application of the dynamic load, the waves were reflected on the model boundaries. To avoid intense reflections due to the perturbations, the energy absorbing boundaries were used at the bottom, left, and right sides of the soil mass.

The software meshed the model into triangles with 6 or 15 nodes during a process called meshing. In PLAXIS software, there are 5 meshing types: very coarse, coarse, medium, fine, and very fine (PLAXIS Manual 2005). The type of meshing depends on the importance of subject. By choosing the very course meshing, the results would not be very accurate, while choosing very fine mesh increases the time of the analysis. As a result, medium meshing was chosen in the software. Also, finer meshing was considered for more sensitive areas such as the tunnel wall; as shown in Fig. 2.

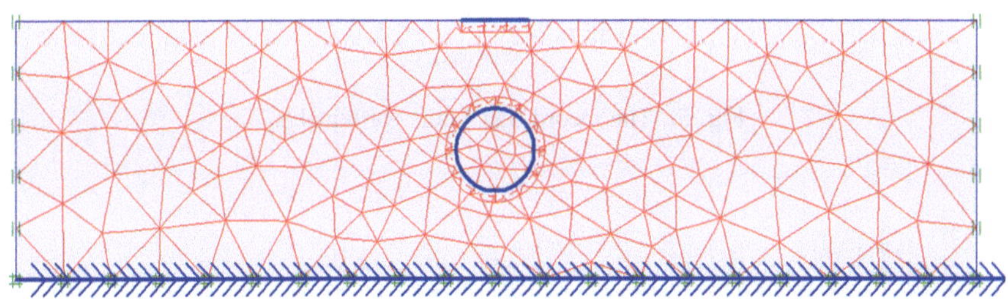

Fig. 2. Boundary conditions and mesh used in numerical simulation.

Table 4 shows the characteristics related to the real earthquake records. In this table, three records were used which were located away from active fault with different maximum accelerations and effective duration of motion. The maximum accelerations for all the described records were scaled to 0.2 g; in the other words, three different records with maximum acceleration amplitude of 0.2 g were separately applied to the mass floor of the soil.

Table 4. Characteristics earthquake records (PEER center).

Row	Record name	PGA (g)	Magnitude (Richter)	Rjb (km)	D9-95 (s)
1	Elcentro	0.31	7.2	18.3	24.1
2	Sanfernando	0.27	6.61	19.33	16.71
3	Sanluis	0.011	6.19	63.34	17.84

The real earthquake records are shown in Figs. 3, 4, and 5. These records were all scaled to the maximum acceleration of 0.2 g (Figs. 6, 7, and 8).

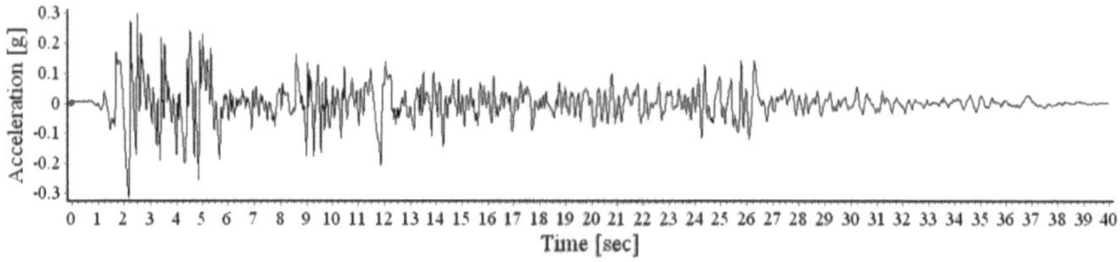

Fig. 3. The record El Centro earthquake.

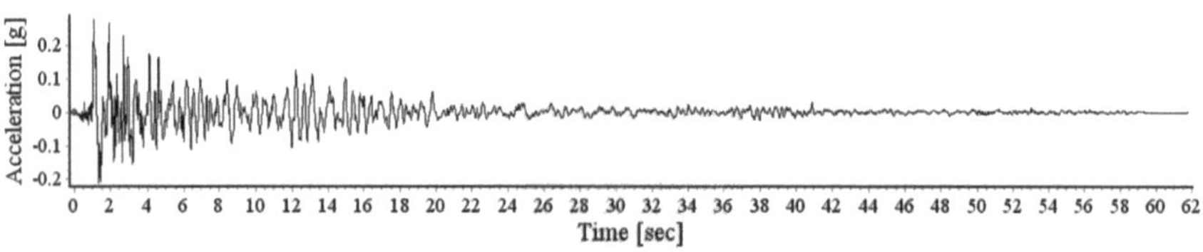

Fig. 4. The record San Fernando earthquake.

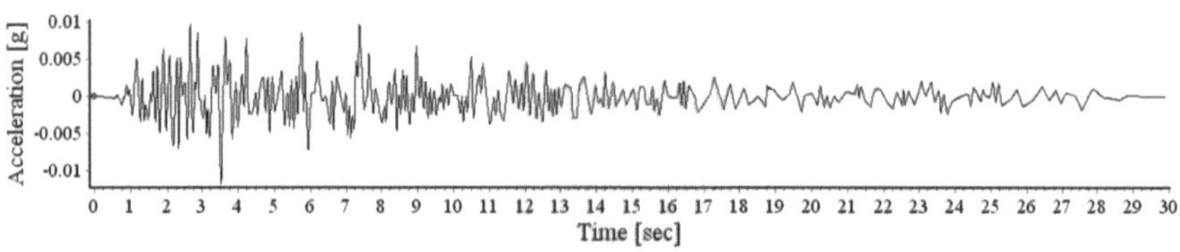

Fig. 5. The record San Louis earthquake.

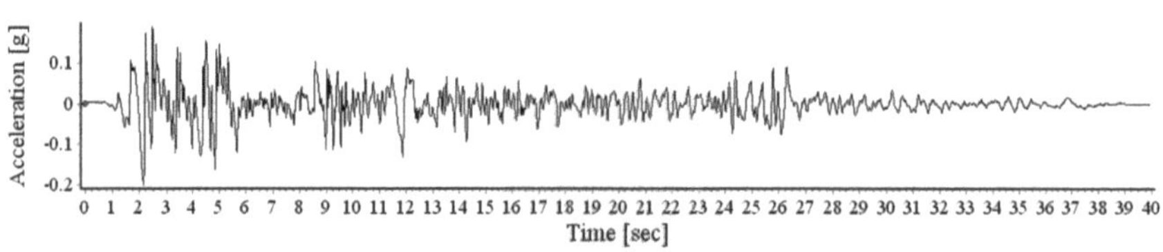

Fig. 6. The record scaled to 0.2 g related to El Centro earthquake.

2 Results and Discussion

At first, the records which were scaled to 0.2 g were applied to the soil mass floor in Plaxis software and, after the dynamic analysis, the soil surface record at point A, located at the ground surface and under the foundation center with coordinates (0 and 50), was obtained. Similar to the tunnel-less case, once again, by excavating a circular

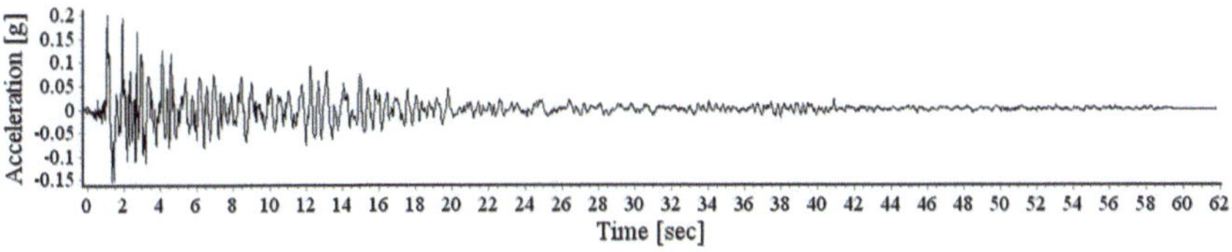

Fig. 7. The record scaled to 0.2 g related to San Fernando earthquake.

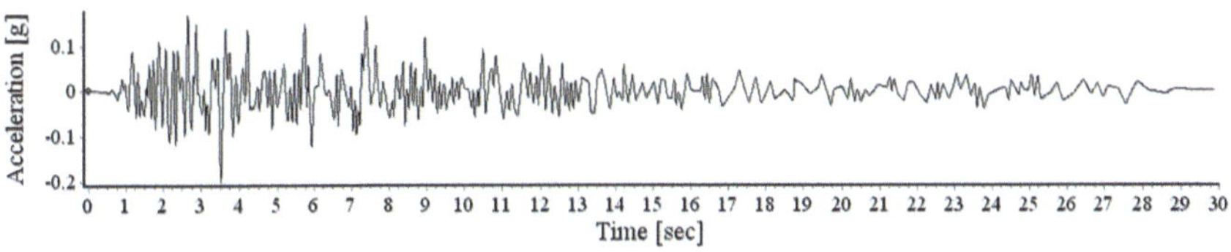

Fig. 8. The record scaled to 0.2 g related to San Louis earthquake.

tunnel in the soil layer, the 0.2 g scaled records were applied to the soil mass floor and the soil surface record was obtained. Finally, the earthquake acceleration at the soil surface, response spectrum, and Fourier spectrum for both cases (with and without tunnel) were compared to each other. In this study, WO indicates the situation without tunnel, while CT means a circular tunnel. Also, EL is El Centro earthquake, SANF denotes San Fernando earthquake, and SANL represents San Louis earthquake.

2.1 Acceleration on the Soil Surface

Figures 9, 10, and 11 show the acceleration of the soil surface for the cases with and without tunnel under different earthquake records. As is clear from the figures, the excavation of tunnel will change the earthquake acceleration.

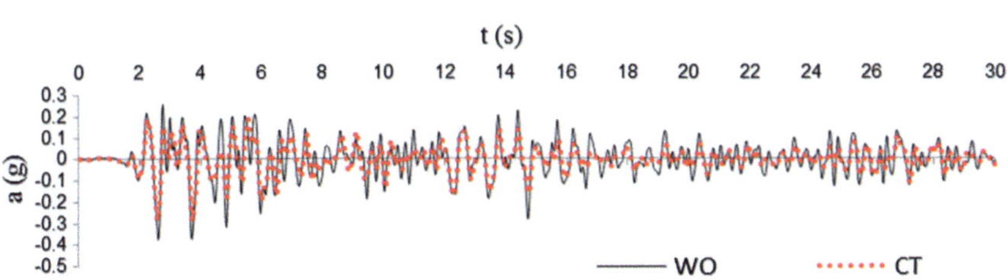

Fig. 9. Acceleration of soil surface for the cases with and without tunnel for El Centro earthquake.

Figure 12 shows the change percent of maximum ground acceleration for the case with tunnel in comparison to the case without tunnel. According to the mentioned figure, circular tunnel decreased the maximum acceleration for all the records, as compared to the case without the tunnel. In this figure, the amount of A is obtained

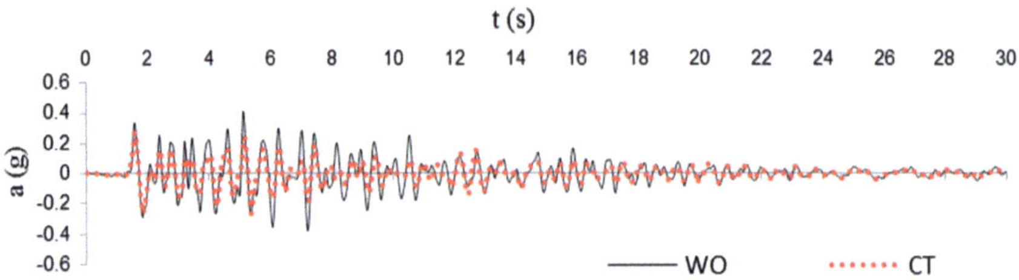

Fig. 10. Acceleration of soil surface for the cases with and without tunnel for San Fernando earthquake.

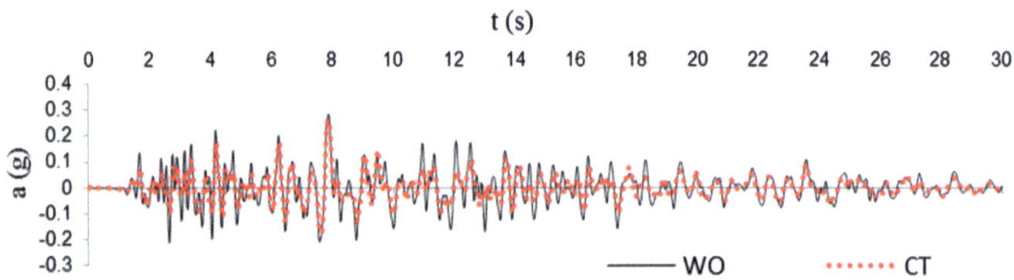

Fig. 11. Acceleration of soil surface for the cases with and without tunnel for San Louis earthquake.

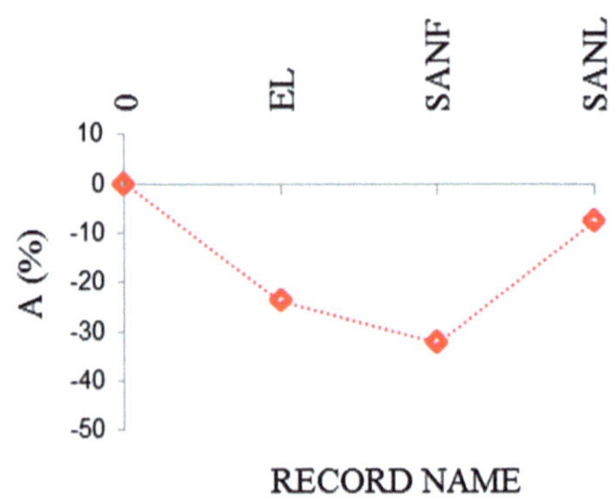

Fig. 12. Variation percent of maximum acceleration.

from Eq. 1 (Variation percent of maximum acceleration), where a_T shows the maximum acceleration of the soil surface for the case with tunnel where, a_{WT} represents the maximum acceleration of the soil surface for the case without tunnel. The positive side of the vertical axis shows the increase percent of the maximum acceleration and its negative side reflects the decrease percent of the maximum acceleration.

The time variation (%) of the maximum acceleration is also shown in Fig. 13. The described figure shows that in El Centro earthquake, time of the maximum acceleration increased due to tunnel excavation, while the time of maximum acceleration decreased in

San Fernando earthquake. But the maximum acceleration time in Saint Louis earthquake did not change for both cases. In this figure, t is obtained from Eq. 2, (time variation percent of maximum acceleration) where t_T and t_{WT} denote the time of the maximum acceleration of the soil surface for the case with tunnel and without tunnel, respectively.

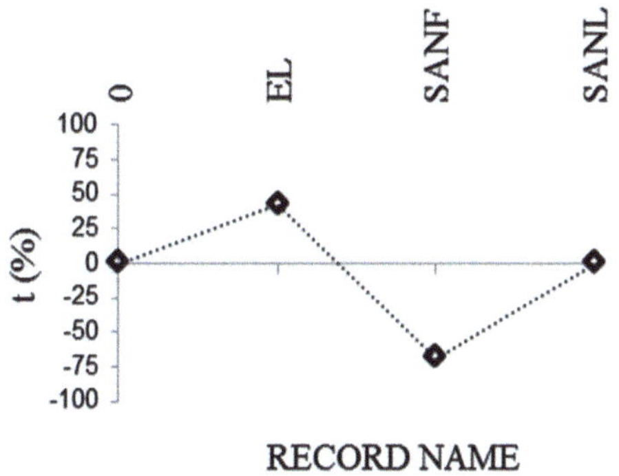

Fig. 13. Time variation percent of maximum acceleration.

The positive and negative sides of the vertical axis show the increase and decrease percent of the maximum acceleration time, respectively.

2.2 Structural Acceleration Spectrum

In both cases (with and without tunnel), the earthquake records were applied to the soil mass floor and record of the soil surface was extracted using the dynamic analysis by PLAXIS software. Records of the soil surface were imported to Seismosignal software and the 1-degree-of-freedom structural acceleration spectrum was drawn for each record.

Figures 14, 15 and 16 show the structural acceleration spectrum for both cases under different earthquake records. As can be observed, excavation of a circular tunnel reduced the structural acceleration spectrum.

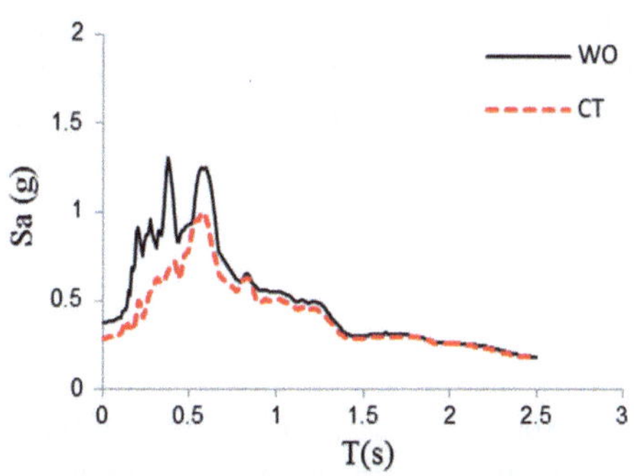

Fig. 14. Structural acceleration spectrum for El Centro earthquake.

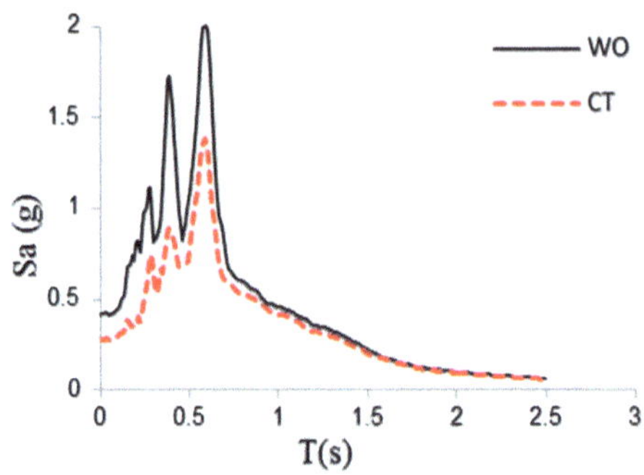

Fig. 15. Structural acceleration spectrum for San Fernando earthquake.

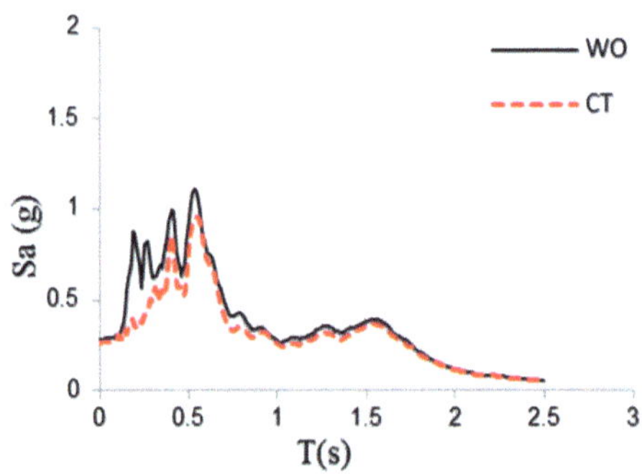

Fig. 16. Structural acceleration spectrum for San Louis earthquake.

2.3 Effective Duration of Motion of the Record

Each acceleration record has an effective duration of motion, which can be defined in different ways, with the energy method being the most important and accurate one. In this method, the time during which 5–95% of the earthquake energy is released, will be measured. Figure 17 shows the changes (%) in the effective duration of record motion for both cases.

According to this figure, in El Centro and San Fernando earthquakes, tunnel excavation increased the effective duration of motion, while in San Louis earthquake, the tunnel reduced the effective duration of motion. In this figure, the amount of d was obtained from Eq. 1 (Variation percent for the effective time of motion):

$$d = \frac{d_T - d_{WT}}{d_{WT}} * 100 \tag{1}$$

In the above equation, d_T and d_{WT} are the effective duration of motion for the cases with and without tunnel, respectively. The positive side of the vertical axis shows the increase percent of the maximum acceleration; while its negative side indicates the decrease in the mentioned value.

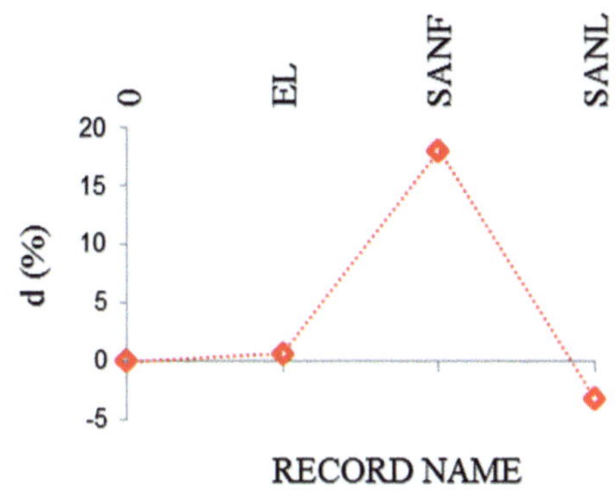

Fig. 17. Variation percent for the effective time of motion.

2.4 Fourier Spectrum for Acceleration of Soil Surface

Figures 18, 19 and 20 show Fourier spectrum of soil surface acceleration for cases under different earthquake records. As it clearly shows, tunnel excavation changed the Fourier spectrum. Figure 21 shows the variations in the maximum amplitude of Fourier spectrum for the with-tunnel case in comparison with the without-tunnel case. According to this figure, excavation of the circular tunnel reduced the maximum amplitude of the Fourier spectrum for all the records in comparison to the without-tunnel case.

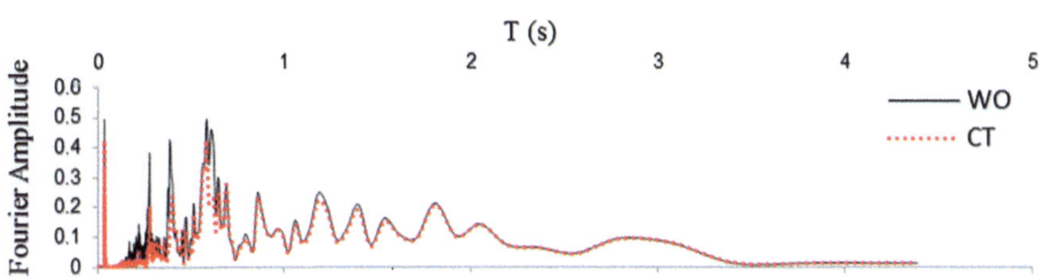

Fig. 18. Comparing the Fourier spectrum of soil surface for the cases with and without tunnel in El Centro earthquake.

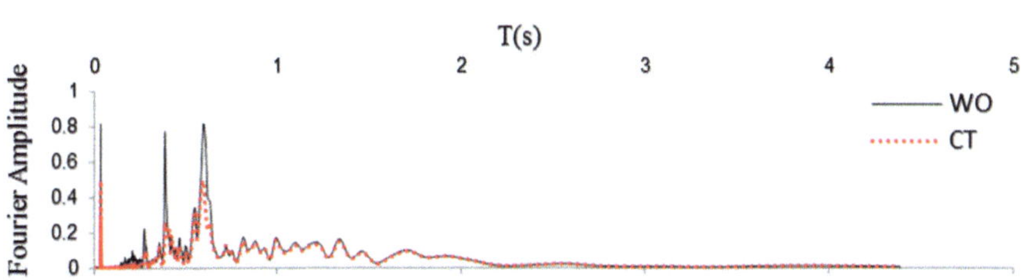

Fig. 19. Comparing the Fourier spectrum of soil surface for the cases with and without tunnel in San Fernando earthquake.

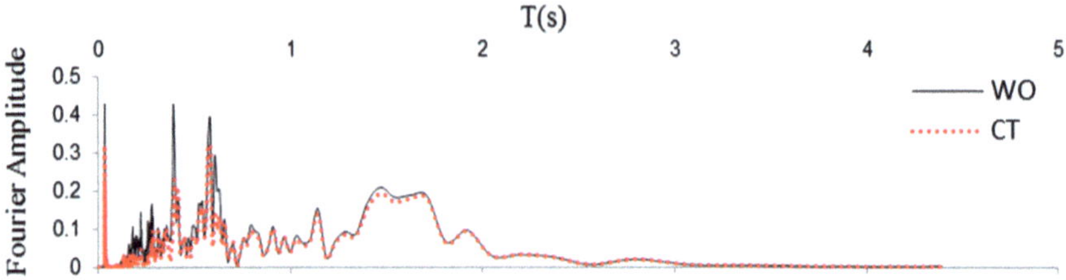

Fig. 20. Comparing the Fourier spectrum of soil surface for the cases with and without tunnel in San Louis earthquake.

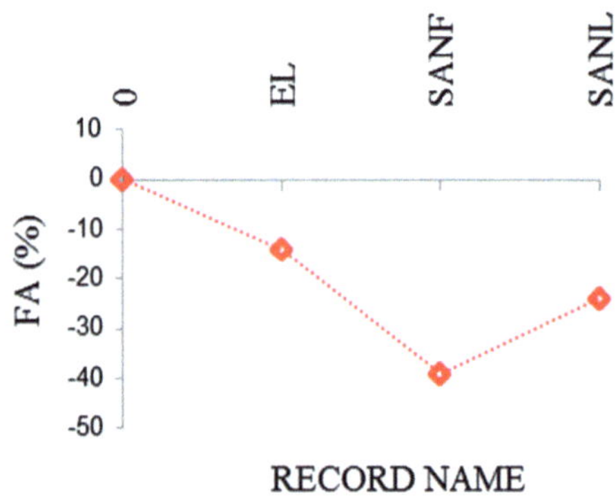

Fig. 21. Variation percent for the maximum Fourier amplitude.

In Fig. 21, FA value was obtained from Eq. 2 (Changes in the maximum Fourier amplitude):

$$FA = \frac{FA_T - FA_{WT}}{FA_{WT}} * 100 \tag{2}$$

where FA_T is the maximum amplitude of spectrum for the case of with tunnel while, is the maximum amplitude of spectrum for the case of without tunnel. The positive side of the vertical axis shows the increase percent of the maximum acceleration whereas its negative side indicates the decrease percent of the maximum acceleration. Figure 22 shows the changes percent in the dominant frequency for the case of with tunnel compared to those of without–tunnel case. In El Centro and San Fernando earthquakes, the dominant frequency for both cases was the same and remained unchanged; but, in Saint Louis earthquake, the dominant frequency reduced. In this figure, the amount of f was obtained from Eq. 3 (changes of dominant frequency):

$$f = \frac{f_T - f_{WT}}{f_{WT}} * 100 \tag{3}$$

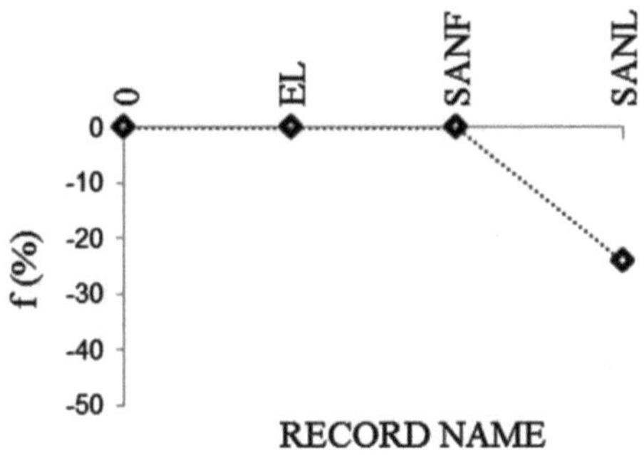

Fig. 22. Variation percent for the dominant frequency.

In which f_T is the dominant frequency for with-tunnel case and f_{WT} denotes the dominant frequency of without-tunnel case. The positive and negative sides of the vertical axis show the increase and decrease percent of time of the maximum acceleration, respectively.

3 Conclusion

In this study, real earthquake records were applied to the soil mass floor for both cases of with and without circular tunnel to extract the earthquake record at the ground surface by PLAXIS software and to study the record characteristics such as acceleration, response spectrum, effective duration of motion, and Fourier spectrum by Seismosignal software. Results of this study indicated that:

1. Excavation of circular tunnels had direct impact on the earthquake ground acceleration so that the maximum acceleration in all the three earthquake records was reduced and time of the earthquake was affected by tunnel excavation, depending on the type of earthquake record.
2. The structural response spectra were different between with- and without-tunnel cases; in a way that, excavation of a circular tunnel reduced the response spectrum as compared to the without-tunnel case.
3. The effective duration of earthquake motion was also affected by the tunnel excavation. Depending on the type of record, this parameter decreased, increased, or remained unchanged.
4. Tunnel excavation also influenced Fourier spectrum of the record. By tunnel excavation, the maximum amplitude of the Fourier spectrum of the records was reduced and the changes in the dominant frequency were different depending on the record type.
5. In general, it can be concluded that assuming constant characteristics for soil and tunnel, tunnel excavation changes the characteristics of the earthquake records at the soil surface. However, the amount of the changes has a direct relationship with earthquake records. By selecting each record, it is possible to observe responses different from other records. Also, in order to construct the superstructures over the tunnels or vice versa, it is necessary to study and analyze the potential risks.

References

Abuhajar, O., El Naggar, H., Newson, T.: Effects of underground structures on amplification of seismic motion for sand with varying density. In: Pan-Am CGS Geotechnical Conference. Ontario, Canada (2011)

Afifipour, M., Sharifzadeh, M., Shahriar, K., Jamshidi, H.: Interaction of twin tunnels and shallow foundation at Zand underpass, Shiraz metro, Iran. Tunn. Undergr. Space Technol. **26** (2), 356–363 (2011). https://doi.org/10.1016/j.tust.2010.11.006

Atkinson, J.H., Potts, D.M.: Subsidence above shallow circular tunnels in soft ground. J. Geotech. Eng. Div. ASCE. **103**(4), 307–325 (1977)

Attewell, P.B., Farmer, I.W.: Ground deformations resulting from shield tunneling in London clay. Can. Geotech. J. **11**(3), 380–395 (1974). https://doi.org/10.1139/t74-039

Baziar, M.H., Moghadam, M.R., Kim, D.S., Choo, Y.W.: Effect of underground tunnel on the ground surface acceleration. Tunn. Undergr. Space Technol. **44**, 10–22 (2014). https://doi.org/10.1016/j.tust.2014.07.004

Besharat, V., Davoodi, M., Jafari, M.K. Effect of underground structures on free-field ground motion during earthquakes. In: 15th World Conference on Earthquake Engineering. Lisbon, Portugal (2012)

Cilingir, U., Madabhushi, S.G.: A model study on the effects of input motion on the seismic behaviour of tunnels. Soil Dyn. Earthq. Eng. **31**(3), 452–462 (2011). https://doi.org/10.1016/j.soildyn.2010.10.004

Clough, G.W., Schmidt, B.: Soft clay engineering, excavation and tunneling (Chapter 8). In: Brand, E.W., Brenner, R.P. (eds.) Elsevier

Dravinski, M.: Ground motion amplification due to elastic inclusions in a half-space. Earthq. Eng. Struct. Dyn. **11**(3), 313–335 (1983). https://doi.org/10.1002/eqe.4290110303

Leach, G.: Pipeline response to tunneling. Unpublished Paper (1985)

Lee, V.W.: Three dimensional diffraction of elastic waves by a spherical cavity in an elastic half-space: closed form solutions. Soil Dyn. Earthq. Eng. **7**(3), 149–161 (1988). https://doi.org/10.1016/S0267-7261(88)80019-8

Mail, R.J., Gunn, M.J., O'Reilly, M.P.: Ground movement around shallow tunnels in soft clay. In: 10th International Conference on Soil Mechanics and Foundation Engineering, pp. 323–328. Stockholm (1983)

Maleki, M., Sereshteh, H., Mousivand, M., Bayat, M.: An equivalent beam model for the analysis of tunnel-building interaction. Tunn. Undergr. Space Technol. **26**(4), 524–533 (2011). https://doi.org/10.1016/j.tust.2011.02.006

Oteo, C. S. & Sagaseta, C. (1996). Some Spanish experiences of measurement and evaluation of ground displacements around urban tunnels. In Geotechnical Aspects of Undergoing Construction in Soft Ground, pp. 731–736. London

Pao, H.Y., Maw, C.C., Achenbach, J.D.: The diffraction of elastic waves and dynamic stress concentrations. J. Appl. Mech. **40**(4) (1973) http://dx.doi.org/10.1115/1.3423178

Peck, R.B.: Deep excavation and tunneling in soft ground. State of the art report. In: Proceedings of the 7th International Conference on Soil Mechanics and Foundation Engineering, pp. 225–290. Mexico (1969)

Pitilakis, K., Tsinidis, G., Leanza, A., Maugeri, M.: Seismic behaviour of circular tunnels accounting for above ground structures interaction effects. Soil Dyn. Earthq. Eng. **67**(1–15) (2014) http://dx.dio.org/10.1016/j.soildyn.2014.08.009

PLAXIS Manual: Finite element code for soil and rock plasticity. Delft University of Technology, Netherlands (2005)

Smerzini, C., Aviles, J., Paolucci, R., Sanchez-Sesma, F.J.: Effect of underground cavities on surface ground motion under SH wave propagation. Earthq. Eng. Struct. Dyn. **38**(12), 1441–1460 (2009). https://doi.org/10.1002/eqe.912

Sun, Ch., Wang, Q.: Effects of underground structure on acceleration response of site. Adv. Mater. Res. **368–373**, 2791–2794 (2012)

Wong, K.C., Shah, A.H., Datta, S.K.: Diffraction of elastic waves in a halfspace. II. Analytical and numerical solutions. Seismol. Soc. Am. **75**, 69–92 (1985)

Yiouta-Mitra, P., Kouretzis, G., Bouckovalas, G., Sofianos, A.: Effect of underground structures in earthquake resistant design of surface structures. ASCE, Dyn. Response Soil Prop. 1–10 (2007). http://dx.doi.org/10.1061/40904(223)16. http://peer.berkeley.edu/

A Case Study of Roadway Embankment Construction Over Existing Sewers in Montreal, Canada

Laifa Cao[1(✉)] and Scott Peaker[2]

[1] Geotechnical Engineer, WSP Canada Inc, Toronto, Canada
Laifa.Cao@wsp.com
[2] Geotechnics, WSP Canada Inc, Toronto, Canada

Abstract. This paper presents a case study of the influence of a roadway embankment on existing sewers with diameter or width of approximately 3.3 to 4.3 m. The proposed embankment would be constructed directly above the existing sewers, which would induce an additional load. Traditional analysis methods cannot account for the structural/soil interaction and the benefit of the lateral supporting characteristics of the soils. In order to evaluate the loading of the proposed embankment on the existing sewers, a structural/soil interaction analysis using a two-dimensional finite element computer program was performed for various loading cases. The sensitivity of soil parameters was also considered in the analysis. From the analysis, the loadings on the sewers including axial force, bending moment and shear force were obtained and then the structural capacity of the existing sewers was checked. Light-weight sand fill or expanded polystyrene (EPS) geofoam was considered to reduce the additional vertical and horizontal loads on the sewers. The EPS geofoam or a structural protection system consisting of a concrete slab supported by concrete piles was considered to eliminate any incremental vertical and horizontal loads on the sewers. These mitigation methods were compared. It is found that the light-weight sand fill is most economic material to reduce the additional loads on the sewers, the EPS geofoam is most suitable material to eliminate incremental loads on the sewers for a lower embankment, and the structural protection system is only option to eliminate incremental loads on the sewers for a higher embankment.

1 Introduction

Redevelopment of roadway systems is a frequent occurrence in urban centers as infrastructure ages and as our reliance on transportation networks increases. A roadway embankment was proposed to replace an existing roadway viaduct in the downtown area of Montreal, Canada. The height of the embankment is up to 9 m. There are numerous challenges associated with roadway embankment construction in urban areas, such as the construction impact on existing utilities, nearby buildings, etc.

This paper presents a case study of the construction influence of a roadway embankment on existing sewers with a diameter or width of approximately 3.3 to 4.3 m. The proposed embankment would be constructed directly above the existing sewers, which would induce an additional load onto the existing sewers. Traditionally,

S. Wang et al. (eds.), *New Solutions for Challenges in Applications of New Materials and Geotechnical Issues*, Sustainable Civil Infrastructures,
https://doi.org/10.1007/978-3-319-95744-9_17

Marston-Spangler theory (Spangler and Handy 1982) is used for the analysis of buried culverts/sewer pipes. However, this method cannot account for the benefit of the lateral supporting characteristics of the embedment soils. In order to evaluate the loading of the proposed embankment on the existing sewers, a soil/structure interaction analysis using a two-dimensional finite element computer program was performed for various loading cases. The sensitivity of soil parameters was also considered in the analysis. From the analysis, the loadings on the sewers, including axial force, bending moment and shear force were obtained and then the structural capacity of the existing sewers was checked. If the existing sewer could not take the additional loads, mitigation methods to reduce or eliminate the additional vertical and horizontal loads on the sewer were considered.

To reduce the additional vertical and horizontal loads, consideration was given to using light-weight sand fill and expanded polystyrene (EPS) geofoam, which are more compressible than conventional fill, to induce positive arching above the sewer. To eliminate any incremental vertical and horizontal loads, consideration was given to use of ultra-lightweight fill (i.e. EPS geofoam) to replace the conventional backfill, or use of a structural protection system consisting of a concrete slab supported by concrete piles to pick up the embankment weight. These methods were compared and the most economical or suitable method was proposed.

2 Site Conditions and Roadway Embankment Construction

The site of the proposed roadway embankment is located in Montreal, Canada. In the study area, a new embankment with an approximate width of up to 70 m and with a height up to approximately 9 m needs to be constructed to replace aging and severely deteriorating roadway viaducts. As the proposed embankment would be constructed directly above the existing sewers, an additional load will be imposed on the existing sewers. These sewers ("Collecteurs") were installed between 1963 and 1990 and were not designed for the additional surcharge. Figure 1 shows a typical cross section of existing sewers and viaducts, along with the proposed new roadway embankment.

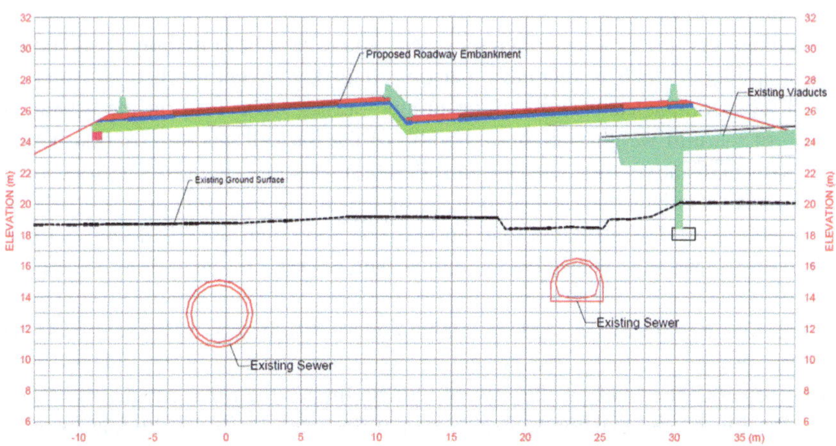

Fig. 1. Existing condition and proposed roadway embankment relative to existing grade

The existing sewers used in the case study includes two types: (1) circular type of concrete pipe with an inside diameter of 3.6 m and thickness of 350 mm; and (2) horseshoe type of concrete pipe with an inside diameter of 2.7 m and thickness of 203 mm to 305 mm.

A geotechnical investigation, including borehole drilling, standard penetration tests (SPT), pressuremeter tests and laboratory testing of soil index parameters, was carried out to support the design of the embankment.

The investigation revealed subsurface soils consisting of: loose to compact fill over very loose to loose sandy silt to silty sand, interbedded with marl, which is in turn underlain by dense to very dense glacial till of sand to sandy silt. Limestone bedrock was encountered at depths of 6.2 to 11.6 m (El. 12.6 to 6.7 m) below the existing ground surface. Piezometric levels were measured at depths of 1.9 to 3.2 m (El. 15.9 to 14.2 m) below the existing ground surface.

Based on the subsurface information encountered in the boreholes, the existing sewers were founded in the dense to very dense glacial till of sand to sandy silt or bedrock.

3 Review of Design Approach for Buried Sewers

The traditional design methods for assessing the capacity of buried sewers are based on the Marston-Spangler theory developed in the 1910's (Spangler and Handy 1982). The buried conduit has been divided into two categories: (1) flexible and (2) rigid. Independent analysis and design approaches have been developed for each. When designing rigid pipes (i.e. concrete pipes), it is customary to assume that the pipe is affected mainly by a vertical pressure caused by soil and traffic; a horizontal reacting pressure is generally negligible. For flexible pipes, the vertical load causes a deflection of the pipe, which in turn results in a horizontal supporting soil pressure. The load on buried conduits consists of earth loads and live loads, both of which depend on the relative stiffness of the soil and conduit. Conduits can be classified as being constructed: (1) in-trench, (2) positive projecting, (3) negative projecting, (4) imperfect trench, and (5) in-tunnel.

It is likely that the existing concrete sewers would have been initially designed for a trench condition, which is relatively narrow ditch dug in undisturbed soil and then covered with compacted backfill to the ground surface. The load on the sewers is a function of the unit weight, height and width of the backfill, and the friction between the backfill and native soil, without consideration of soil cohesion. The load on the rigid pipe is less than the total weight of trench above the conduit due to the soil friction. Based on the calculated load on the pipe and determined bedding factor, the required pipe three-edge bearing strength was then selected for an appropriate factor of safety.

After construction of the roadway embankment, resulting in additional grade raise above the original, the negative projecting condition needs to be considered for the existing sewers. The load on the sewers will be higher than that for the trench condition. Obviously, the existing sewers were not designed for the additional load.

Although the traditional design methods have been the standard practice since the early 1900's, research and field monitoring have confirmed that such approaches are overly conservative due to assumptions associated with limited knowledge on soil-pipe interaction, especially the inability to account for the benefit of the lateral supporting characteristics of the embedment soils (Smeltzer and Daigle 2005). Research on soil-structure interaction with the application of the finite element (FEM) method over the past thirty years has led to the development of an enhanced design method (i.e. a direct design method) for rigid concrete pipe and its embedment systems. The direct design method determines the actual moments, thrust and shears in the buried pipe. Furthermore the direct design method based on the FEM analysis enables the evaluation of various installation conditions (time-history) as part of the design analysis. A standard installation design method was developed by the American Concrete Pipe Association (ACPA 1998) based on a large number of FEM model simulations and field studies of four standard installations. The American Society of Civil Engineers (ASCE) adopted a standard installation design method through ASCE 15-93 (ASCE/ANSI 1993). Canadian Standards Association (CSA) includes a standard installation design method for the design and installation of buried structures in the Canadian Highway Bridge Design Code (CSA S6-06 2006).

This project required the load on the existing sewers to be reviewed based on CSA S6-06 (2006). The earth load determined from the unit weight of overfill soil over the top of the sewer and its effect were determined by an analysis of soil-structure interaction based on the characteristics of the installation based on CSA S6-06 (2006).

For a circular concrete pipe, the total vertical earth load on a buried pipe ranges from 1.35 to 1.45 times the weight of the column of earth applied over the outside diameter of pipe; the total horizontal earth load on a buried pipe ranges from 0.3 to 0.45 times the weight of the column of earth over the outside diameter of pipe in standard installations with various bedding thickness and soils; and the earth pressure distribution on the pipe can be calculated based on the force diagram provided in CSA S6-06 (2006) for four standard installations.

Since the installation method for the existing sewers is not clear, it is not applicable to use the Marston-Spangler theory or the force diagram provided in CSA S6-06 (2006) to review the load condition of the existing sewers under the new embankment. Use of FEM analysis to simulate the time-history of construction is the best way to determine the loading conditions on the existing sewers under the new embankment as well as for the modelling of protection systems to mitigate impacts on the existing sewers.

4 Finite Element Analysis

A soil-structure interaction analysis using a two-dimensional FEM computer program (RS2, Version 9) was performed to evaluate the loads on the existing circular sewer and the horseshoe-shaped sewer.

4.1 Circular Sewer

The soil profiles selected for the analyses were based on the boreholes drilled near the sewer. The boundary elevations for each soil layer used in the analysis are provided in Table 1. The surface of bedrock was taken at El. 10.7 m. Bedrock is modelled as an incompressible material and thus is considered as the fixed boundary. The marl was assumed to be replaced by new fill during embankment construction. Note that the unit weight of new fill is much higher than that of marl. This assumption will give a higher loading on the sewer, which is a conservative approach. The soil surrounding the sewer is assumed to be backfill comprised of existing fill and the thickness of the existing fill is taken as 0.5 m. The new pavement on the top of the embankment consists of 290 mm asphalt concrete, 200 mm granular base of MG20, and 590 mm granular subbase of MG112. The groundwater level was taken at El. 15.8 m in the analysis.

Table 1. Unfactored soil, EPS Geofoam and concrete parameters

Material type (Elevation)	Unit weight γ (kN/m^3)	Friction angle (degree)	Young's modulus (MPa)	Poisson's ratio
New fill	18.2	32	50	0.3
New light weight sand fill	16.8	32	30	0.3
Granular fill (MG-20)	22.9	35	100	0.3
Granular fill (MG-112)	18.9	32	50	0.3
MSE fill	22.0	36	50	0.3
Existing fill	18.2	32	50	0.3
Marl (>El. 17.0 m)	12.1	–	6.9	0.3
Silt to silty sand (till) (El. 17.0 to 10.7 m)	20.9	36	100	0.3
EPS Geofoam	0.21	–	5	0.12
Concrete/Asphalt	24	–	29700	0.2
Water	9.81	–	<0.2	0.49

Since the existing circular, reinforced concrete sewer installed in 1990 was found to be in good condition on the basis of visual inspection and intrusive testing, consideration was given to reduce the weight of the new embankment using relatively light weight sand fill above the existing sewer with the secondary intention of inducing the arching effect. The objective, of course is to limit the load on the sewer so as not exceed its structural capacity for axial force, bending moment and shear force. Consideration is also given to use compressible material (i.e. EPS geofoam) to achieve this same effect. Since the EPS geofoam will compresses more than the surrounding fill, the load on the sewer will be less than the calculated overburden pressure due to the side friction, resulting in a positive arching above the sewer. This is called the induced trench method (also called an imperfect ditch). Field measurements conducted by Vaslestad et al. (2011) confirmed the positive arching effect on buried rigid culverts using the EPS geofoam.

The unfactored soil parameters and elevations of the soil layers used in the analysis are listed Table 1, including unit weight (γ), Young's modulus (E), Poisson's ratio (μ), and angle of internal friction (ϕ). Young's modulus for cohesionless soils was estimated from the available pressuremeter data. The coefficient of earth pressure at rest, K_o is taken as 0.5. The cohesion (undrained shear strength) of the marl is taken as 40 kPa. The apparent cohesion of the EPS geofoam is taken as 68 kPa. The cohesion for other soils is taken as 0.1 kPa. The assumed E for the EPS geofoam is taken as recommended for EPS100 by the Transportation Research Board (2004) and the Poisson's ratio for the EPS geofoam is as recommended by the EPS Industry Alliance (2012). It is noted that the analysis result is not significantly affected by Young's modulus of EPS geofoam. The assumed Young's modulus for the soils is based on CSA S6.1-14 (2014).

The vertical earth pressure (VEP) is calculated from the unit weight of soils, γ and the lateral earth pressure (LEP) is calculated from K_o. Thus the load factor is related to γ and K_o in the finite element analysis. At the serviceability limit states (SLS), the load factor for γ is taken as 1.0 for γ and K_o. At the ultimate limit states (ULS), the load factor for γ is taken as 1.0 or 1.375 and the load factor for the LEP is taken as 0.8 or 1.0 for different cases based on CSA S6-06 (2006). As the LEP is calculated from the VEL and K_o, the value of K_o is taken as 0.29 or 0.5 as summarized in Table 2.

Table 2 Ranges of load factor and coefficient of earth pressure at rest

Load combination (LC)	α_D	α_E	α_L	α_W	α_{LE}	K_o
LC1 for SLS	1.0	1.0	0.9	1.0	1.0	0.5
LC1 for ULS	1.1	1.375	1.87	1.1	0.8	0.29
LC2 for ULS	1.0	1.0	1.0	1.0	1.0	0.5

The analysed circular sewer at this section is a reinforced concrete pipe with inside diameter of 3.6 m and wall thickness of 349 mm. The sewer was modelled as beam elements. The dead load (DL) of the sewer is calculated from γ, taken as 24 kN/m^3 and its Young's modulus is taken as 29730 MPa. The moment of inertia for the sewer is taken as 0.00354 m^4 at the SLS and as 0.0025 m^4 in the ULS. The load factor for concrete γ is taken as 1.0 or 1.1 based on CSA S6-06 (2006).

The sewer was assumed as being full of fluid with the properties of water. The unit weight of water is taken as 9.81 kN/m^3 and its Young's modulus is taken as a value of less than 200 kPa to make the vertical and horizontal water pressure (WP) on the sewer wall equivalent to the hydrostatic pressure.

The traffic live load (LL) is taken as 17.6 kPa. The load factor for the LL is taken as 0.9 at the SLS and 1.87 at the ULS.

The soils were modeled as Mohr-Coulomb elastoplastic materials and the concrete and steel were modeled as elastic materials.

The load combination (LC) used in the analysis is as follows

$$LC = \alpha_D DL + \alpha_E VEP + \alpha_L LL + \alpha_W WP + \alpha_{LE} LEP \tag{1}$$

where α_D is the load factor for the dead load of sewer; α_E is the load factor for vertical earth pressure; α_L is the load factor for traffic live load; α_W is the load factor for water pressure within the sewer; and α_{LE} is the load factor for lateral earth pressure. The values of the load factor are listed in Table 2. Two load combinations at the ULS and one load combination at the SLS have be analyzed. The other load combinations do not govern and thus were not checked.

The construction sequence was simulated in the FEM static analysis as follows:

Stage 1 Initial ground condition with the sewer and existing retaining wall;
Stage 2 Installation of the EPS geofoam or placement of the light weight fill;
Stage 3 Construction of embankment; and
Stage 4 Adding LL surcharge.

Figure 2 shows the partial FEM mesh after construction of the embankment with a 4.7 m wide and 1 m high zone of the EPS geofoam above the existing sewer for the SLS load case at a section of up to 9 m high embankment. The finite element mesh was 92 m long and 16.3 m high. The bottom boundary was restrained from both vertical and horizontal movements. The left-hand and right-hand boundaries were free to move in the vertical direction.

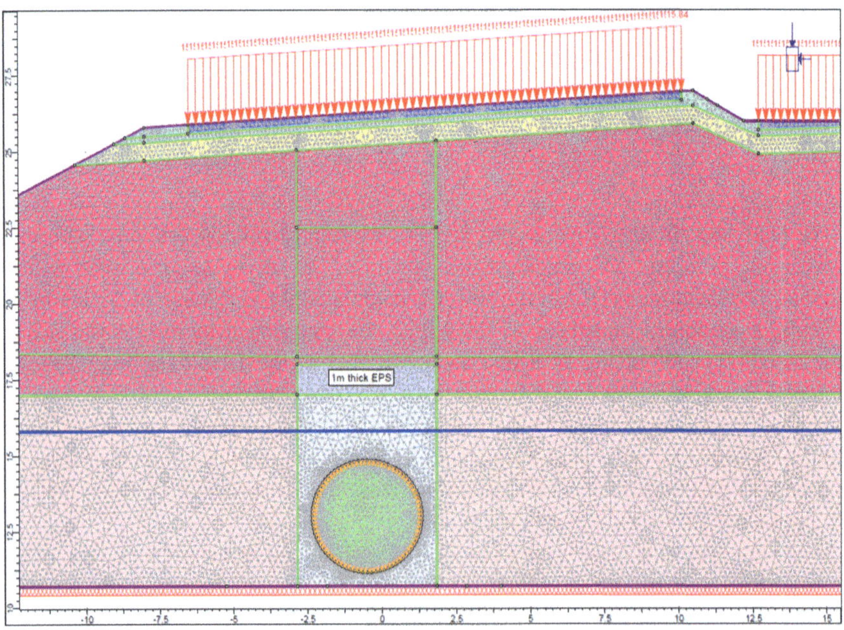

Fig. 2. FEM mesh after embankment construction with EPS geofoam

Typical computer output, such as total vertical stress and bending moment for the existing sewer after the embankment construction with 1 m thick EPS geofoam (Stage 4) for the SLS load case, is shown in Fig. 3. It clearly shows that the total vertical stress on the top of the sewer is significantly reduced to only about 60% of the total vertical overburden pressure due to the positive arching effect. This result is consistent with the field measurements by Vaslestad et al. (2011). This phenomenon could not be modelled by the traditional Marston-Spangler theory.

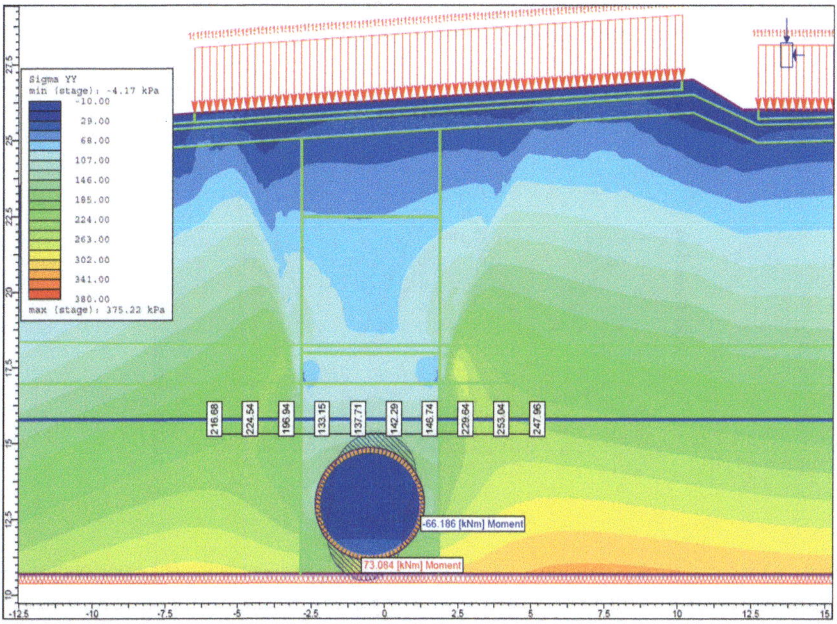

Fig. 3. Vertical stress in soil and bending moment for existing sewer after embankment construction with EPS geofoam

Figures 4, 5 and 6 compare the axial force, bending moment and shear force for the existing sewer after the embankment construction, respectively for three different options: (1) conventional fill; (2) light weight sand fill in a zone of 4.7 m width by 5.5 m height above the sewer; (3) EPS geofoam in a zone of 4.7 m width by 1 m height above the sewer. The axial force, bending moment and shear force for the existing sewer at the existing condition are also shown in these figures. It is found that the axial force, bending moment and shear force for the existing sewer after the embankment construction increase due to the new embankment construction. Comparing to the conventional fill option, both the light weight fill and EPS geofoam options can reduce the additional axial force, bending moment and shear force mainly due to the the positive arching effect. Since the value of Young's Modulus for the EPS geofoam is much smaller than that of the light weight sand fill, the reduction in the axial force, bending moment and shear force using 1 m thick EPS geofoam is more significant than that achieved using the 5.5 m thick light weight sand fill. It is noted that separate stages for the installation of the EPS geofoam (or the light weight sand fill) following embankment construction must be used in the FEM analysis to simulate the positive arching effect. If the EPS geofoam or light weight sand fill following embankment construction are modelled in one single stage, the positive arching effect is insignificant. The variation of soil parameters was also considered in the analysis. It is found that the influence in the variation of the soil parameters to the analysis results is insignificant as the variation in the load factors is bigger than that of the soil parameters themselves.

Based on the results of FEM analysis for various loading cases, it is found that the existing sewer can accommodate the loads induced by the new embankment with a

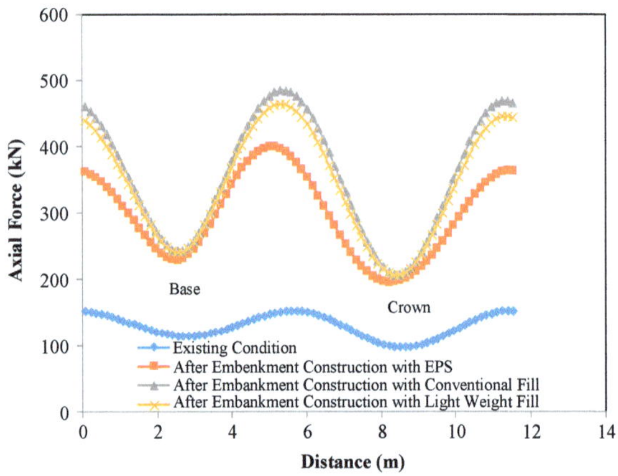

Fig. 4. Comparison of axial force for existing sewer after embankment construction with existing condition for mitigation options

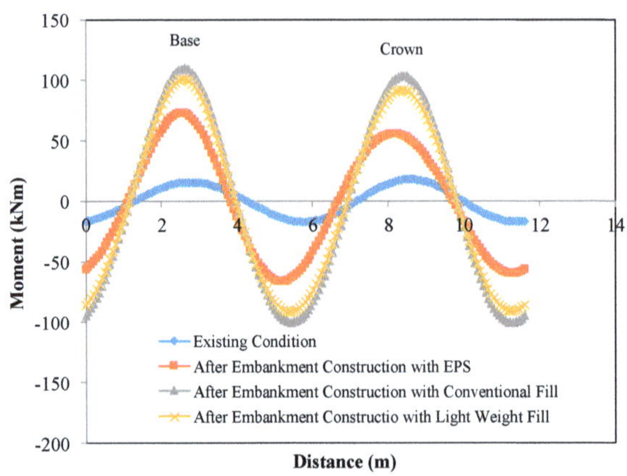

Fig. 5. Comparison of bending moment for existing sewer after embankment construction versus existing condition for different mitigation options

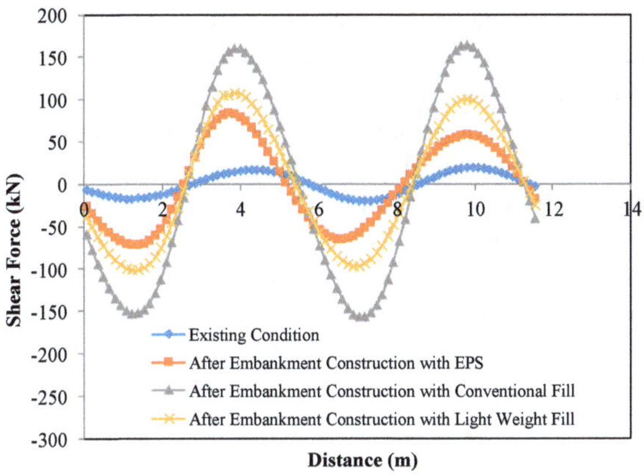

Fig. 6. Comparison of shear force for existing sewer after embankment construction versus existing condition for different mitigation options

zone of light weight sand fill. Thus the light weight sand fill option, which is more economical than the EPS option, is adopted for the new embankment construction over the existing circular sewer.

4.2 Horseshoe-Shape Sewer

The horseshoe-shaped concrete sewer was constructed in 1963. Based on visual condition surveys and core drill findings it was considered that this ageing sewer could not sustain any additional load. Thus two options were proposed: (1) a protection system ("relieving platform") consisting of a concrete slab supported by concrete piles to relieve the additional loading on the existing sewer; and (2) using ultra-light-weight fill (i.e. EPS geofoam) to construct the new embankment.

The soil profiles and parameters used in the analysis are the same as those used for the circular sewer.

The equivalent inside diameter of this sewer is 2.7 m. The thickness of sewer wall varies from 203 mm at the top, 305 mm at the base to 381 mm at the sides. The average thickness and moment of inertia for the sewer were used in the analysis. The moment of inertia at the ULS was taken as 40% of that at the SLS. Young's modulus for the sewer was taken as 20,000 MPa based on the testing of core samples taken from the sewer.

The load cases conducted in the FEM analysis are the same as those for the circular sewer, except that the sewer was assumed to have a dry interior (without water), which is a more critical case.

Figure 7 shows the partial FEM mesh with a relieving platform consisting of a 10.5 m wide by 0.8 m thick concrete slab supported by 406 mm diameter piles at a spacing of 3 m center to center for a new embankment height of 7.5 m for the SLS load case. A gap of 200 mm was placed beneath the concrete slab and the underlying soil to ensure that the embankment load would be taken up by the pile foundation. Even through the width of the relieving platform is about 3 times the sewer width, it could not totally relieve the additional lateral earth pressure on the sewer. With the relieving platform installed 1 m below the existing ground surface, there is a net reduction in the vertical load on the existing sewer and the resulting maximum axial force, bending moment and shear force are not greater than those in the existing condition. Figure 8 shows the typical computer output of total vertical stress and axial force in the existing sewer for the SLS case. It clearly shows that there is no increment in the total vertical stress surrounding the existing sewer.

Figure 9 shows the bending moment for the existing sewer prior to and after a 7.5 m high new embankment construction with the protection system for the SLS load

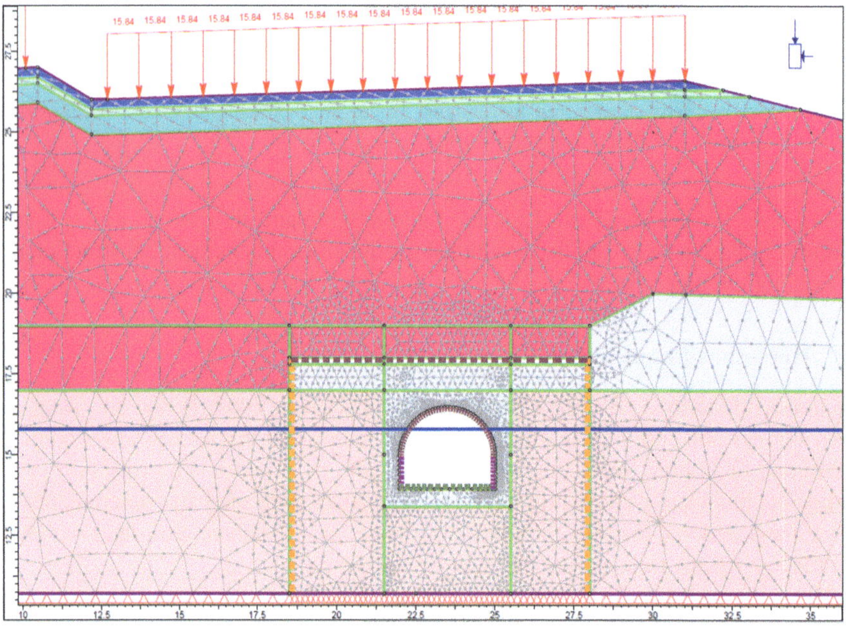

Fig. 7. FEM mesh after embankment construction with a protection system

case. It is found that the maximum bending moment after the embankment construction is not more than that in the existing condition although its distributions are not the same due to slight increase in the horizontal stress and slight decrease in the vertical stress. Similar trends are also found for the axial and shear forces in the existing sewer. The bending moments for the existing sewer after the embankment construction with the EPS option are also shown in Fig. 9 for comparison. Although the EPS geofoam thickness is the same as that of the new embankment and the width of EPS geofoam is 3 times the sewer width, there is an increment of bending moment for the sewer base for a 7.5 m high new embankment as the EPS option could not fully eliminate the increase of the lateral load near the sewer base.

Figure 10 shows the bending moment for the existing sewer prior to and after the embankment construction with the protection system and EPS geofoam for a 3 m high

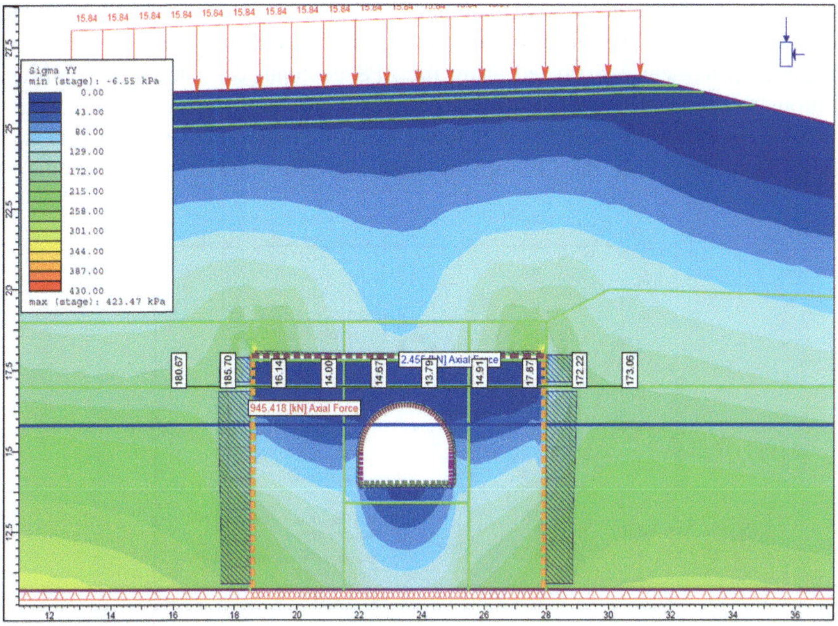

Fig. 8. Vertical stress in soil and axial force in structures after embankment construction with a protection system

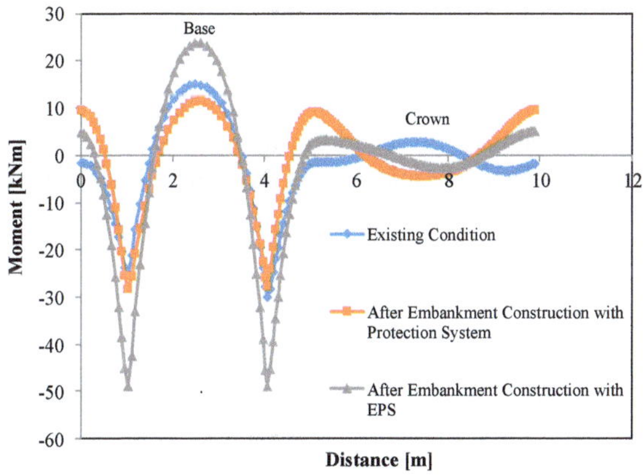

Fig. 9. Comparison of bending moment for existing sewer prior to and after embankment construction with a protection system and EPS geofoam for a 7.5 m high embankment

new embankment for the SLS load case. Similar to the 7.5 m high embankment, the maximum bending moment after the embankment construction with the protection system is not more than that in the existing condition. For the new embankment construction with the EPS geofoam, there is no increment in the bending moment for the sewer base as shown in Fig. 10. The slight decrease in the bending moment for the sewer crown is due to the positive arching effect as the modulus of the EPS geofoam is less than that of soil. There is also no increment in the axial and shear force, indicating that the embankment constructed using the EPS geofoam can successfully eliminate additional loads on the existing sewer for the 3 m high new embankment.

Based on a series of analyses, it is found that for a higher new embankment with a height greater than 3 m, the EPS option could not fully eliminate the additional horizontal load at the sewer base. Thus a more expensive approach using a concrete slab supported by concrete piles is recommended to relief the additional loading on the existing sewer. For a lower new embankment with a height not more than 3 m, the EPS option, which will not increase the bending moment, shear and axial forces for the existing sewer and is more economical than the concrete slab protection system, is recommend.

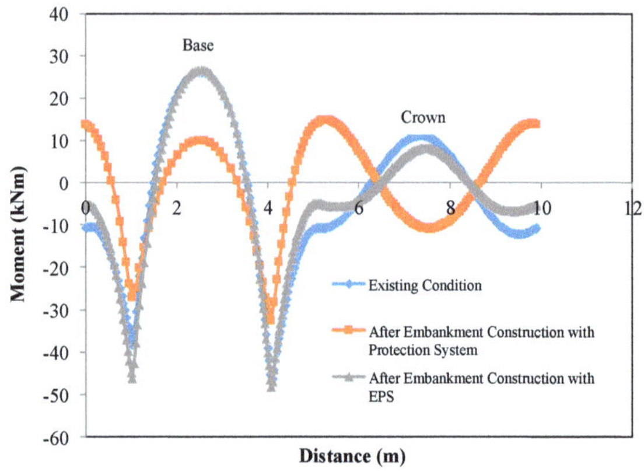

Fig. 10. Comparison of bending moment for existing sewer prior to and after embankment construction with a protection system and EPS geofoam for a 3 m high embankment

5 Conclusions

The additional load on the existing sewers due to new embankment construction can be successfully reduced or eliminated using mitigation methods such as light-weight sand fill, EPS geofoam, or protection systems consisting of a concrete slab supported by concrete piles. Traditional design methods for buried culverts and rigid pipes cannot be used to estimate the loading condition of the existing sewers under the effects of the new embankment due to their lack of ability to account for soil-structure interaction. The widespread use of the finite element method in geotechnical applications makes it possible to directly analyze the interaction of the buried culverts and soil under various mitigation scenarios.

It is found that the light-weight sand fill is most economic material to reduce the additional loads on the sewer, the EPS geofoam is most suitable material to eliminate incremental loads on the sewer for a lower embankment, and the structural protection system is only option to eliminate incremental loads on the sewer for a higher embankment.

References

ASCE/ANSI: Standard Practice for Direct Design of Buried Precast Concrete Standard Installation (SIDD). ANSI/ASCE 15–93. American Society of Civil Engineers/American National Standards Institute (1993)

ACPA: Concrete Pipe Handbook. American Concrete Pipe Association, Irving, Texas, US (1998)

CSA: Canadian Highway Bridge Design Code, CAN/CSA S6-06. Canadian Standards Association, CSA International, Toronto (2006)

CSA: Commentary on CSA S6-14, Canadian Highway Bridge Design Code, CAN/CSA S6.1-14. Canadian Standards Association, CSA International, Toronto (2014)

Spangler, M.G., Handy, R.L.: Soil Engineering, 4th edn. Harper & Row, New York (1982)

Smeltzer. P.D., Daigle, L.: Field performance of a concrete pipe culvert installed using standard installations. In: 33rd Annual general conference of the canadian society for civil engineering. Toronto, Canada (2005)

The EPS Industry Alliance: Expanded Polystyrene (EPS) Geofoam Applications & Technical Data, Crofton, MD (2012)

Transportation Research Board: Guideline and Recommended Standard for Geofoam Applications in Highway Embankments, NCHRP Report 529, Washington, D.C (2004)

Valestad, J., Sayd, M.S., Johansen, T.H., Wiman, L.: Load reduction and arching on buried rigid culverts using EPS Geofoam. Design methods and instrumented field tests. In: 4th International Conference on the Use of Geofoam Blocks in Construction Applications, Norway (2011)

Correction to: Ground Motion Amplification Induced by Shallow Circular Tunnel in Soft Soil

Mohsen Mousivand, Hesam Aminpour, and Nura Ebrahimi

Correction to: Chapter "Ground Motion Amplification Induced by Shallow Circular Tunnel in Soft Soil" in: S. Wang et al. (eds.), *New Solutions for Challenges in Applications of New Materials and Geotechnical Issues*, Sustainable Civil Infrastructures, https://doi.org/10.1007/978-3-319-95744-9_16

The original version of the chapter was published without the co-authors' names Hesam Aminpour and Nura Ebrahimi. The authors' names have been updated.

The updated version of this chapter can be found at https://doi.org/10.1007/978-3-319-95744-9_16

S. Wang et al. (eds.), *New Solutions for Challenges in Applications of New Materials and Geotechnical Issues*, Sustainable Civil Infrastructures,
https://doi.org/10.1007/978-3-319-95744-9_18

Author Index

S. Wang et al. (eds.), *New Solutions for Challenges in Applications of New Materials and Geotechnical Issues*, Sustainable Civil Infrastructures,
https://doi.org/10.1007/978-3-319-95744-9

MIX
Papier aus verantwortungsvollen Quellen
Paper from responsible sources
FSC® C105338

If you have any concerns about our products,
you can contact us on
ProductSafety@springernature.com

In case Publisher is established outside the EU,
the EU authorized representative is:
Springer Nature Customer Service Center GmbH
Europaplatz 3, 69115 Heidelberg, Germany

Printed by Libri Plureos GmbH
in Hamburg, Germany